Synthese Library

Studies in Epistemology, Logic, Methodology, and Philosophy of Science

Volume 529

The aim of *Synthese Library* is to provide a forum for the best current work in the methodology and philosophy of science and in epistemology, all broadly understood. A wide variety of different approaches have traditionally been represented in the Library, and every effort is made to maintain this variety, not for its own sake, but because we believe that there are many fruitful and illuminating approaches to the philosophy of science and related disciplines.

Special attention is paid to methodological studies which illustrate the interplay of empirical and philosophical viewpoints and to contributions to the formal (logical, set-theoretical, mathematical, information-theoretical, decision-theoretical, etc.) methodology of empirical sciences. Likewise, the applications of logical methods to epistemology as well as philosophically and methodologically relevant studies in logic are strongly encouraged. The emphasis on logic will be tempered by interest in the psychological, historical, and sociological aspects of science. In addition to monographs *Synthese Library* publishes thematically unified anthologies and edited volumes with a well-defined topical focus inside the aim and scope of the book series. The contributions in the volumes are expected to be focused and structurally organized in accordance with the central theme(s), and should be tied together by an extensive editorial introduction or set of introductions if the volume is divided into parts. An extensive bibliography and index are mandatory.

Arthur C. Schwaninger

Cognitive Metaphysics

How the Predictive Brain Constructs Reality

Springer

Arthur C. Schwaninger
University of Zurich
Zürich, Switzerland

ISSN 0166-6991 ISSN 2542-8292 (electronic)
Synthese Library
ISBN 978-3-032-05419-7 ISBN 978-3-032-05420-3 (eBook)
https://doi.org/10.1007/978-3-032-05420-3

This Springer imprint is published by the registered company Springer Nature Switzerland AG
The registered company address is: Gewerbestrasse 11, 6330 Cham, Switzerland

In memory of Jürg and Napo

Contents

Part III Constructing the Manifest World

Acronyms

FEP	Free Energy Principle
IGP	Information Grouping Principle
L/LN	Binary/Three-Valued Logic
LGN	Lateral Geniculate Nucleus
PP	Predictive Processing
PPC	Primacy of Physics Constraint
QM	Quantum Mechanics
RPP	Robust Predictive Process
UTM	Universal Turing Machine
V_1	Primary Visual Cortex

Chapter 1
Introduction

When we ordinarily encounter the world, we find it to be *pre-divided* into different objects of various pre-given kinds. For instance, we find the world to consist of living and non-living things. Looking more closely at the realm of living things, we find that some living things are in certain respects more similar to one another than they are to other organisms, such as in their appearance, behaviour, or genetic code. Instinctively, we group these beings by introducing certain categories in our language, such as 'dog', and then refer to particular objects in the world accordingly. It appears that in some way my neighbour's chihuahua and the dog Lassie have a lot in common and *go together* by the natural order of things. Likewise, it seems that other objects we find in the world, such as electrons, tables and numbers, naturally do not belong in this group.

Much of scientific research is engaged in drawing proper distinctions to explain certain phenomena or to make the best possible predictions. A physicist can, for instance, inform us that neutrons and electrons are of distinct kinds, but that a clear-cut distinction between the spatial and the temporal cannot be made in light of Einstein's theory of relativity. A botanist might develop a taxonomy that classifies walnuts as fruits and imply that they are in some way more closely related to apples than to carrots.

While we ordinarily, or in a scientific context, identify relevant kinds and draw appropriate distinctions, philosophers who engage in metaphysics attempt to draw those distinctions and identify those kinds that are the most basic or *fundamental* to reality. Just like scientists, metaphysicians develop models. These are not, however, specific to any particular domain of inquiry but concern all possible domains and are of the most general nature, aiming to describe the universe as a whole. For example, the seventeenth-century philosophers Baruch Spinoza, René Descartes, and Gottfried Wilhelm Leibniz believed that all of reality consists of substances of various kinds as ultimate bearers of properties, though they differed in their views about the nature of these substances. For Spinoza, God was the only substance

A. C. Schwaninger, *Cognitive Metaphysics*, Synthese Library 529,
https://doi.org/10.1007/978-3-032-05420-3_1

throughout the cosmos, and all of nature is part of this eternal, infinite, self-caused one. Descartes, however, believed that in addition to the infinite, divine substance, reality divides itself into two other kinds, namely minds and matter. Leibniz, on the other hand, conceived reality as an endless pool of extensionless minds or mind-like substances that he calls *monads*. For all three philosophers, metaphysics is an attempt to formulate a general description of the world; it is the quest for a general theory of how things, in the broadest sense of the term, relate to one another.[1]

The metaphysical enterprise conceived in this form is often associated with Plato's metaphor of 'carving nature at its joints'. In ancient Greece, philosophers hoped to secure knowledge that holds universally and would not be swept away by the flux of everyday change. Plato's strategy was to locate the objective fault lines of the world first, identifying the correct set of categories that make up the world as it is. These categories divide the world in the way that joints of a skeleton divide bones. Only by having these categories at our disposal can we formulate universally true statements about reality.[2] In contemporary terms, this project is often described as the search for the fundamental metaphysical structure of reality—a framework of joints, kinds, and patterns that best captures how the world is independently of how we think or talk about it (see, e.g., Sider, 2011, ch. 1).[3]

While many relevant categories are discussed in metaphysics, such as time, free will, numbers, and causation, the primary concern of this book is how we must think about the material world. In particular, the book focuses on the role of ordinary objects such as trees and tables in the greater context of reality, and how we ought to think about their nature in contrast to the subatomic world. One of the important distinctions philosophers make when discussing the fundamental structure of the world is the distinction between mind-dependent and mind-independent entities. This categorial distinction places every thing into one of two buckets: either into the bucket of things that exist in dependence on our perceptual and mental activity, or into the bucket of those that exist independently of us. This distinction is vital to understanding the fabric of reality, as it separates things like the Bogeyman from things that are actually found in the world. However, the history of philosophy shows that it is not a straightforward matter into which bucket ordinary objects fall, nor how these buckets ought to be defined in the first place. During the eighteenth and nineteenth centuries, *idealism*—the view that the material objects of our experiences are dependent on the mind—gained significant acceptance within intellectual thought. This enthusiasm faded with the rise of the analytic tradition in the early twentieth century. Today's mainstream view remains strongly opposed

[1] See Moore (2012) who offers a detailed description of how the metaphysical enterprise evolved over time and how our understanding of what it means to make general sense of the world has changed.

[2] See Plato, *Phaedrus*, trans. R. Waterfield (2002), pp. 265–266.

[3] The terms 'world', 'universe', and 'reality' are used interchangeably, unless stated otherwise and refer to the set of all facts and things that exist, as discussed later in the book.

to idealist tendencies, and academics largely assume that the world is, in structural terms, much as we experience it to be.

From our ordinary experiences, we tend to believe that tables and chairs have a life of their own: 'ordinary perceptual experience seems to provide us at every moment with crowds of mind independent things' (Campbell & Cassam, 2014, p. 29). This shapes our ordinary beliefs about ordinary objects in a metaphysical sense. Our common sense suggests that ordinary, everyday objects are entities that exist independently of the human mind, and that the essence of their nature is entirely reducible to what lies in the world, rather than being attributable to the observer. This common sense also appears to prevail over time. As the Irish philosopher and clergyman George Berkeley already wrote in the eighteenth century:

> It is indeed an opinion strangely prevailing amongst men, that houses, mountains, rivers, and in a word all sensible objects have an existence natural or real, distinct from their being perceived by the understanding. (Berkeley, 1710, p. 524)

Numerous philosophers have adopted this common-sense view. While our perception might be limited, incomplete, or sometimes misguided, most aspects of the material world we ordinarily perceive are taken to be aspects of the world as it really is. For the philosopher John Searle, for instance, ordinary objects play a crucial part in what constitutes reality and should be taken to be the things as we ordinarily think of them:

> Well, now look at the objects around you, the chairs and tables, houses and trees. These objects are not in any sense "subjective." They exist entirely independent of whether or not they are experienced. (Searle, 2004, p. 3)

In a discussion about tables and cats, Bertrand Russell writes that there is no simpler and better alternative assumption 'than the common-sense hypothesis that there really are objects independent of us, whose action on us causes our sensations' (Russell, 1912, p. 10). And as Daniel Korman simply puts it: 'ordinary objects are mind-independent' (Korman, 2015, p. 94), meaning that they have their properties independently of our observing them. For many, the mind-independent existence of ordinary objects is a Moorean fact—a fact that is more certain than any alternative.

Despite it seeming evident that ordinary objects exist in some mind-independent way, there are various concerns we should have about making such claims. Consider L. Frank Baum's children's classic, *The Wonderful Wizard of Oz*, published in 1900, in which Dorothy travels along a yellow road to Emerald City, where she hopes to receive the wizard's help in returning home. Anyone who seeks to enter the city is told to wear a pair of spectacles that protect their eyes from the city's 'brightness and glory' (p. 117). In the city, Dorothy sees the 'beautiful houses all built of green marble and studded everywhere with sparkling emeralds' (p. 121). She eventually learns, however, that the self-proclaimed wizard possesses no magical powers whatsoever and that the city is not as beautifully green as expected. The city only appears green to its visitors and inhabitants because they are all forced to wear green-tinted spectacles at all times. Had they been allowed to take off the glasses, the city would not have appeared greener than any other city.[4]

When people enter Emerald City with their tinted glasses, they are misled into believing that objects in the environment have certain properties due to the glasses they are wearing, and are unable to see the world as it really is. They think that all the objects in the city share the common property of being made of emerald—a property that distinguishes them from objects in other cities. Thus, they believe that there is a categorial distinction to be made in the world that separates their city from others. However, this distinction, which is based on the division implicitly suggested by the glasses and the procedure of entering the city, does not seem to carve reality at its joints.

But how could the philosophers of Emerald City know that they are not using the correct categories and distinctions to talk about the material world without taking off their glasses? The concern that we might not have access to the structure that truly carves up reality has led many philosophers to question both the general ambitions of this intellectual endeavour and the precise methodology by which metaphysics has been conducted at various stages in history.

In recent years, a new theoretical framework for understanding the functioning of the brain has emerged under the label of *predictive processing*. According to this framework, the brain is actively 'hallucinating' the world of our experiences. The approach aligns with recent developments in Bayesian modelling across various disciplines and is closely related to the rise of generative neural networks in artificial intelligence and machine learning. While computer scientists have developed models that automatically generate short pieces of text indistinguishable from human writing, as well as images that can easily be mistaken for the work of famous artists (Shahriar, 2022), brain scientists have found evidence that conceptually and computationally similar models may explain visual perception. For this reason, a small number of researchers—including neuroscientists and evolutionary biologists—have raised concerns that our human representation and conceptualisation of the world are largely mistaken, arguing that we must draw certain idealist metaphysical conclusions from these novel insights. These researchers do not merely suggest that humans fail to see the world *as it is* due to epistemic limitations; rather, they claim that the human cognitive system *actively* imposes mental structures onto the world, thereby misleading us in our thinking about it.

Our cognitive system motivates us to believe that the world possesses a certain structure composed of discrete objects, and that concepts such as *object*, *property*, *individual*, and *causation* are particularly apt for capturing this structure. In reality, however, all these ideas and concepts may be nothing more than neural constructs. Rather than guiding us toward an accurate understanding of reality, they might in fact be misleading us in our efforts to comprehend how the world truly is. According to this line of thought, what we ordinarily perceive is not a mind-independent world,

[4] Others like Pippin (2005, p. 182) and Leahey (2017, p. 169) draw a similar analogy between *The Wonderful Wizard of Oz* and Kantian ideas on perception, as discussed over the course of this book.

but rather a neural construct akin to the illusions produced by the spectacles worn in Emerald City, which may not be veridical at all. On this view, the brain creates a biologically *useful*, rather than *veridical*, representation of the world. It is said to be *actively constructing* what we experience in a manner that is not truth-preserving. Given these considerations, it seems reasonable that some have suggested we may need to revise the very concepts employed in scientific and philosophical discourse when attempting to speak about reality itself. The idealist tradition, in the broadest sense of the term, has gained new momentum from the very domain that once seemed to undermine it—the natural sciences. While the physical sciences have long been the primary focus of philosophers concerned with the structure of reality, it is now findings from the cognitive sciences that compel us to reconsider our realist conception of the world.

This book investigates the 'spectacles' that we, as humans, are unable to take off—namely, the human cognitive system. It explores how these spectacles shape our perspective on the world from the viewpoint of predictive processing, and what implications this has for how we ought to think about reality. The project is divided into three parts. The first part introduces the problems we face when attempting to develop a metaphysics of the material world and proposes a neuroscience-oriented approach to address them. It is argued that metaphysics ought not to focus primarily on questions of existence, but rather clarify how the various structures we encounter—both in the sciences and in experience—relate to one another in a unified framework. To this end, a form of naturalised Kantian idealism is proposed. The second part presents the neuroscientific foundations necessary for this approach to succeed. It does so by outlining key milestones in the history of neuroscience. We will see that much of early twentieth-century neuroscience took the world as given, assuming that the primary task of our cognitive system is to gather sufficient information to reconstruct an accurate internal model of external reality. In contrast, contemporary research suggests that the brain actively imposes its own structure on incoming information to a far greater degree than classical models assumed. On this basis, the third part examines how our cognitive system ultimately generates the world we perceive, given the limited data it receives, and how this affects the way we must think about the material world. Traditional metaphysical puzzles concerning material reality will be revisited and shown to be addressable from this revised perspective.

The book aims to invite both philosophers and neuroscientists into a shared conversation. While some sections may seem elementary to one group or the other, I trust that the broader exploration will resonate with both. My hope is to offer an accessible introduction to an alternative way of contemplating reality—one that intertwines classical and contemporary metaphysical thought with the evolving landscape of computational neuroscience. By bridging these fields, this work aspires to foster a deeper understanding of both the mind and the world it perceives—or rather, constructs.

Part I
Naturalising Kantian Metaphysics

Chapter 2
Two Worlds in Conflict

2.1 Revolting Miracles

In the biblical story found in 2 Kings 6:1–7, a band of young prophets approached their leader, Elisha, to propose that they gather some wood to build a bigger dwelling.

> They went to the Jordan and began to cut down trees. As one of them was cutting down a tree, the iron axhead fell into the water. "Oh no, my lord!" he cried out. "It was borrowed!" The man of God asked, "Where did it fall?" When he showed him the place, Elisha cut a stick and threw it there, and made the iron float. "Lift it out," he said. Then the man reached out his hand and took it.

Instead of becoming lost at the bottom of the Jordan River, to their amazement the iron axe head miraculously floated back to the surface and could be retrieved. For those present, the event was nothing short of miraculous, a testament to divine intervention. The iron, naturally destined to sink, defied its inherent properties, demonstrating the presence and power of the divine in their midst. The event apparently impressed the scribes and chroniclers enough to deem it worthwhile documenting, ensuring that future generations would remember the extraordinary occurrence. In an era when miracles were seen as concrete proof of divine favour, the story of the floating axe head served as compelling evidence of the supernatural.

As we traverse into our modern era, the lens through which many of us in Western societies view such events has significantly shifted and now makes us find the story of the floating axe head quaint and absurd. It appears particularly antiquated, as unlike other biblical verses rich with symbolic meaning, this story seems to offer little to no dimension of interpretation, but merely reports an occurrence that appears highly implausible.

If today somebody were to report on an iron object floating for inexplicable reasons, our first instinct would be to seek a rational explanation grounded in the principles of science. Rather than announcing the occurrence of a miracle, a team of scientific researchers would likely spring into action. They would begin

A. C. Schwaninger, *Cognitive Metaphysics*, Synthese Library 529,
https://doi.org/10.1007/978-3-032-05420-3_2

measuring the water's surface tension, temperature, and any unusual chemical compositions. Sophisticated instruments would be used to analyse the water for anomalies that could contribute to the iron's buoyancy. A team of material science experts would examine the iron's density, composition, and structural integrity. Using high-resolution imaging technologies, they would rule out any hidden cavities or materials that might cause it to float and try to detect internal features invisible to the naked eye, supported by detailed computer simulations along the way. The possibility of a miracle is, by today's mindset, not even considered an option. Instead, we approach such exceptional events with scientific inquiry, and if all efforts to explain and recreate the event were to fail, we would simply begin to doubt the initial reports.

Ludwig Wittgenstein captures this paradigm shift as follows:

> We all know what in ordinary life would be called a miracle. It obviously is simply an event the like of which we have never yet seen. Now suppose such an event happened. Take the case that one of you suddenly grew a lion's head and began to roar. Certainly that would be as extraordinary a thing as I can imagine. Now whenever we should have recovered from our surprise, what I would suggest would be to fetch a doctor and have the case scientifically investigated and if it were not for hurting him I would have him vivisected. And where would the miracle have got to? For it is clear that when we look at it in this way everything miraculous has disappeared; unless what we mean by this term is merely that a fact has not yet been explained by science which again means that we have hitherto failed to group this fact with others in a scientific system. This shows that it is absurd to say "Science has proved that there are no miracles." The truth is that the scientific way of looking at a fact is not the way to look at it as a miracle. (Wittgenstein, 1965, pp. 10–11)

If something 'supernatural' were to happen—such as a person growing a lion's head or an axe head floating on water—we would investigate it scientifically to understand the underlying causal mechanisms that led to the phenomenon's occurrence. From a scientific perspective, then, 'everything miraculous has disappeared.'

This does not, however, imply that we only accept answers that are intuitively accessible to us. On the contrary, we often place our trust in the scientific enterprise so deeply that we accept its conclusions even when they defy our ordinary intuitions. Ever since the scientific revolution, which began at the end of the Renaissance and culminated in the work of Isaac Newton, it has become clear that the sciences—favouring empirical investigation and quantitative, rational analysis over intuitive reasoning—offer a method far superior to earlier approaches when it comes to predicting natural phenomena. The success of giving common sense little authority is already evident in Copernicus' calculations of the orbits of celestial bodies, which replaced an ideologically biased, geocentric worldview with a radically new, non-anthropocentric cosmic structure. Another example is Newtonian mechanics, which postulates that material bodies remain in motion unless acted upon by an external force—challenging earlier theories that bodies naturally come to rest, as one might intuitively expect. We trust the scientific approach not only because of its explanatory and predictive power, but also because our societies have been able to build extraordinary technologies based on its findings.

2.2 Understanding Reality Through Science

This general trust in science also manifests itself in philosophical inquiry. When Bertrand Russell was once asked about the difference between philosophy and science, he replied:

> Well, roughly, you'd say science is what we know and philosophy is what we don't know. That's a simple definition and for that reason questions are perpetually passing over from philosophy into science as knowledge advances. ... [A]ll sorts of questions that used to be labelled philosophy are no longer so labelled.[1]

The better we understand what the world is like from a scientific perspective, the less we appear to depend on philosophical inquiry. Accordingly, the passing of questions into the sciences also occurs in metaphysics, despite its traditional character as a field independent of empirical findings. Philosophers once constructed theories about reality purely from the armchair, relying solely on deductive logic and introspection. Based on such a non-empirical approach, a philosopher like Spinoza, for instance, would formulate a theory of the universe as bound by a single, infinite substance that connects all things, with mind and matter as two aspects of this unified reality.

Despite its intellectual rigour, this type of highly speculative metaphysical work is more akin to an art form than a genuine quest for understanding. It contrasts sharply with standard contemporary metaphysical thought, which holds that the only reasonable metaphysical assessment of reality comes from interpreting what the sciences reveal about it. To understand how philosophy arrived at this point, we must understand the question that brought it there. A central concern that occupied philosophical thought was the question of how we can distinguish the real from the unreal. For most contemporary philosophers, an object is said to be *real* if it belongs to *reality*, meaning that it is one of 'all things that exist'. Whether or not a type of object exists is the question of ontology. Since ontological work requires some understanding of how this question can be approached, a separate domain called 'meta-ontology' is often introduced to address questions such as what it means for something to be real or to exist.[2] In ordinary language, the notion of existence appears in different forms. We utter statements like 'There exists ice on Mars' or 'Socrates might not have existed', in which we make positive or negative claims about the structure of the universe. As often occurs in the sciences, we may initially believe that things of a certain kind exist, only to revise these beliefs in light of new evidence. For instance, according to the seventeenth-century phlogiston theory, combustible bodies were thought to burn because they contained a fire-like element called phlogiston. Eventually, it turned out that phlogiston does not exist,

---

[1] Bertrand Russell, interview with Woodrow Wyatt, *Bertrand Russell Discusses Mankind's Future* (1960).

[2] Whether this distinction is useful—and thereby implies that ontology can be sensibly conducted as a research domain independent of meta-ontology—is questionable, as the question of existence always requires some meta-ontological preliminary thoughts.

leading instead to the discovery of oxygen, which offered a far better explanation of combustion. Within this scientific context, it makes perfect sense to say, 'Phlogiston does not exist, whereas oxygen does', and we all understand what is meant by such a statement.

In philosophy, the notion of existence is, however, much more puzzling than one might expect from ordinary parlance. Philosophers noticed that when we allow for negative existence statements, such as those concerning fictional characters or refuted scientific entities, we run into a semantic problem. What is a statement like 'unicorns do not exist' about, if unicorns do not exist? One would expect that a statement must be about something rather than nothing. But if unicorns are nothing, then the statement would be simply empty. Thus, it seems the statement must be about something—namely, unicorns. Suggesting that an object does not exist appears to presuppose its existence. As a response to this problem, the philosopher Alexius Meinong (1904) distinguishes between *being* and *existing*, and suggests that there *are* both tables and unicorns, but only the former *exists*. Existence is thereby conceived as a property that objects may or may not possess, like any other property such as being red or being spherical. Accordingly, for every set of properties, there is, as Meinong claims, a corresponding object that either exists or does not exist.

The view of treating existence as a property and of distinguishing it from *being* seems at first appealing. However, upon closer investigation it suffers from various difficulties, making it largely unacceptable to most of the philosophical community. An unintuitive consequence of Meinong's theory is that if every set of properties corresponds to an object, whether existing or not, then there is also an object that only has the property of being blue and no further properties such as size, shape, or extension. This appears to be at odds with our ordinary understanding of being, as when we say 'If an object is round, it is extended'. A true inconsistency arises when we imagine an object that is both round and square at the same time. Such an object appears to involve a contradiction, and this is philosophically unacceptable (Russell, 1905).[3]

Treating existence as a predicate had already been criticised by both Gottlob Frege and Immanuel Kant. The latter regarded this assumption as the root of unsound arguments attempting to prove the existence of God. In Anselm's classical ontological argument, the greatest thinkable being is said to exist by necessity, since its non-existence would diminish its greatness. Thus, if the greatest thinkable being did not exist, it would not be the greatest thinkable being—which leads to a contradiction. Kant's diagnosis of this 'proof' is that it mistakenly treats existence as a real predicate.

So how, then, should we think about existence? The standard modern approach to answering existence questions is often attributed to Quine, who reintroduced metaphysics to mainstream academic philosophy after it had been rejected as a field of study by the positivists (Ney, 2012). Following Bertrand Russell (1905), Willard Van Orman Quine (1948) argues that we misinterpret a statement such as

[3] A detailed discussion on these points is found in Bigaj (2012) and Reicher (2019).

'The present King of France does not exist' because it has not been translated into a suitable structure for analysis. For example, the sentence contains the descriptive expression 'the present King of France', which on its own does not refer to anything, contrary to what we might naively assume. It can, however, be translated into a form that clarifies the referential relationship between the object and the description: 'There is an x such that x is the present King of France'. We call this the *quantificational* structure, as the variable x is a placeholder ranging over a domain of objects, and it determines the truth or falsity of the statement. In this case, since there are no such objects in the world, the statement is true—there is no King of France. The original sentence appears philosophically problematic only because, in its original form, it disguises the underlying quantificational structure. Once formulated correctly, the problem dissolves.

Instead of treating existence as a predicate, this insight leads to a logical formalism that expresses existence using the so-called *existential quantifier*, denoted by the symbol $\exists$. This allows us to rewrite the earlier example as either '$\neg\exists x(Kx)$', meaning 'There is no object satisfying the description of being the King of France', or as '$\forall x(\neg Kx)$', meaning 'For all objects in the world, none satisfies the description of being the King of France'. For readers with a background in the sciences, it is worth noting that this treatment of existence is not unique to philosophy but also finds expression in mathematical logic. In mathematics, objects with the same properties are described as elements of the same set, and yet no mathematician uses a set to express the property of existence. For instance, when a mathematician writes a statement such as 'there exists a real number such that ...', they do *not* write '$x \in \mathbb{E} \cap \mathbb{R} : (\ldots)$', but rather '$\exists x \in \mathbb{R} : (\ldots)$', thereby expressing existence via the quantifier $\exists$ rather than invoking a set $\mathbb{E}$ of all existing things.

Having provided a formalism that avoids the puzzle of non-existent objects, Quine argues that we can now give meaningful answers to questions about whether objects of a certain kind K exist (e.g., objects like tables or electrons), thereby expressing his ontological approach outlined earlier. Of course, such assessments cannot proceed without empirical grounding. Quine suggests that we examine a domain of discourse that has established itself on the basis of empirical investigation and identify its ontological commitments in order to make the best possible judgement about what is real. More precisely, we must first determine which empirical domain we trust the most. For Quine, this is not ordinary perceptual experience, but the sciences—in particular, fundamental physics. Within this scientific domain, certain statements or laws are established. By paraphrasing and reformulating such statements, we can translate the claims made by the chosen science into a logical structure. Quine's famous slogan 'to be is to be the value of a bound variable' captures the idea that our best estimate of what exists is found by looking at which variables a theory quantifies over when it is cast into the appropriate logical form. For instance, when a physical theory describes changes in electrical currents and these laws refer to electrons, we are thereby committed to the existence of electrons—at least insofar as the theory remains unrefuted. As another example, Quine emphasises the existence of sets because every physical theory we currently know depends on mathematics, which in turn presupposes a numerical apparatus built on the concept of sets. Thus, the indispensability of sets in our best scientific theories commits us to their existence (Quine, 1970). By contrast, unicorns play no role in our scientific discourse and are therefore considered non-existent.

This method has formed the foundation of what is now often referred to as the *Quinean* approach to metaphysics. According to this view, there is little distinction between metaphysics and ontology, and the principal task of the metaphysician is to translate scientific discourse into a formal structure from which we can infer which entities exist and which do not. This approach is said to offer a principled way to assess which components of our ordinary experiences are real and which can be discarded (van Inwagen, 1998). The approach makes implicit meta-ontological assumptions, including that existence is not an activity (in contrast to, for instance, Heidegger's notion of existence), and that it is unified and perspective-independent. However, one must be careful about what is commonly referred to as the Quinean approach is not necessarily the only or the correct way to understand Quine's metaphysics. Rather, it rests on a very specific interpretation of Quine's work, most influentially promoted by Peter van Inwagen (1998), and is largely based on an isolated reading of Quine's (1948) monumental article 'On What There Is'.[4]

In summary, then, the Quinean approach suggests that our best attempt at formulating a theory of reality involves little more than consulting the sciences to define a comprehensive list—the *Book of the World*—of all the things they claim to exist. As scientific knowledge evolves, so too must our ontology, and we must be prepared to adapt our ontological commitments when necessary. Should new discoveries invalidate earlier scientific beliefs or introduce novel entities, it is both reasonable and necessary to revise our ontological inventory to reflect these changes.

To the science-minded reader, this approach may seem entirely reasonable. After all, what could have greater authority than the sciences in making sense of reality? This mindset reflects the prevailing zeitgeist, which elevates scientific knowledge and its methods above all other ways of knowing, as it has consistently demonstrated its ability to offer more robust explanations than those grounded in miracles or supernatural postulations.

2.3 The Scientific and the Manifest Image

While a science-based approach to reality may provide one useful perspective on how we conceptualise the world, it can also appear incomplete. Consider a group of scientists tasked with investigating the performance of a symphony orchestra. Their goal is to analyse the musical performance entirely through scientific observation. They measure the vibrations of the strings, the air pressure changes in the brass instruments, and the timing of the percussion. They collect

[4] A concise overview of the Quinean approach can be found in Egerton (2016) and Schaffer (2009).

data on frequencies, decibel levels, and harmonic overtones, and construct detailed models of the acoustics in the concert hall. Equipped with this data, they present their findings, assuming they have fully captured the nature of the performance. However, something remains unaccounted for. Despite the completeness of their physical measurements, they have not addressed the subjective dimension of the experience—such as the emotional impact of a crescendo or the mood evoked by a minor chord. Their scientific approach overlooks how the music is actually perceived and experienced by those in the audience.[5]

The limitations of the scientific approach become even more apparent in light of normative questions. Imagine that in the audience there are two newspaper critics. One of them, a staunch traditionalist with a preference for classical precision, finds the performance unsatisfactory. He criticises the conductor's deviations from the score, the violinist's improvisations, and the brass section's pronounced dynamics, which he sees as excessive and disruptive. His review focuses on what he considers a lack of discipline and respect for the original composition. The other critic, by contrast, responds positively. With a strong appreciation for innovation and emotional expression, she highlights the same features as strengths.

Which of the two critics is right in their judgement? No matter how precisely scientists chart the acoustic details, they will never be able to adjudicate whose review correctly reflects the performance. The critics' opinions are based on personal taste, cultural values, and individual perspectives on musical excellence. These aspects might be *describable* in empirical terms, but they are not *assessable* by them. Our normative judgements about what is good or bad, beautiful or ugly, are not part of scientific reality.

What we find is a profound tension between the scientific worldview and our ordinary common-sense perspective. In modern philosophy, this tension was most famously elaborated by Wilfrid Sellars. He argues that there is a discrepancy between our common-sense understanding of the world and what science tells us the world is *actually* like. We ordinarily think of the world as a place in which we act as agents who are generally free in our decisions, living in a society shaped by norms and values. This is what Sellars calls the *manifest image* of the world, a conception grounded in our human constitution and practical needs. But we can also view the world through the lens of the physical sciences, which describe reality in terms of particles governed by mathematical laws. This is the so-called *scientific image* of the world, a world devoid of meaning, normativity, or agency. According to Sellars, these two images *clash* with one another: one is shaped by our experience as human beings, the other by attempts to describe reality as it is independently of us.

Science does not merely explain how some events we encounter occur as consequences of other natural phenomena. It also includes human beings as part

[5] This example is reminiscent of the 'Mary the Colour Scientist' thought experiment, which explores the limits of scientific knowledge in understanding subjective experiences (see Jackson, 1982).

of the natural order. The sciences reduce human and cultural activities to the natural realm by offering causal explanations of how our behaviour is shaped by physical processes. It is not guided by an immortal, transcendent soul but governed by the same laws that apply to all matter. Human action, at first approximation, is the result of neural firing patterns in the brain. Neuroscience explains how sensory input causes electrical signals to be transmitted through the nervous system, leading to the contraction of muscles and the production of bodily movement. Physics goes even deeper, revealing that every part of our body is composed of matter that behaves according to fundamental laws—the same laws that govern the smallest subatomic particles and the largest galactic structures.

A science-oriented worldview, when confronted with events like the floating axe head, instinctively seeks material explanations such as unusual water properties or hidden mechanisms, and this results in a broader tendency to interpret all phenomena through the lens of physical causality. By prioritising natural explanations over any others, science fosters a belief system in which everything is ultimately reduced to interactions of matter and energy—a worldview referred to as *materialism*:

> The universe, that vast assemblage of every thing that exists, presents only matter and motion: the whole offers to our contemplation nothing but an immense, an uninterrupted succession of causes and effects. (d'Holbach, 1770, p. 15)

This view leads not only to the rejection of any supernatural entity but also to the denial of all normative judgements, such as what is right and wrong, or what is good and evil. In the materialist view, little is left of a *self* and human agency; instead, everything is described in terms of matter standing in some causal relation to other matter. A person with morals, values, and identity turns out to be nothing more than a hunk of matter consisting of flesh and blood.

The sciences confront us with the materialist worldview, which challenges us in our most fundamental beliefs. If nature is to be understood mechanistically, and humans are a part of nature, how then must we conceive human subjectivity? And if there is nothing like a self or human agency, what happens to our free will? How must we think about morality or self-determination? Science offers us an ontology that is far removed from everyday life and yet claims to account for all the facts in the world. It accounts for everything and nothing at the same time. Materialism directly conflicts with many of our ordinary beliefs about identity and agency, reducing all aspects of human life to the physical properties of mere hunks of matter floating in space. However, living in an age of scientific reverence, many of us are willing to accept this image.

The significance of the tension between the manifest and the scientific image cannot be overstated; it may well represent the deepest and most enduring conundrum within modern philosophy. In nearly every philosophical problem we encounter, this tension surfaces in one form or another. When philosophers debate the existence of free will, they must navigate the chasm between these images, confronting empirical evidence that suggests unconscious neural processes precede conscious decision-making.[6] When philosophers delve into the nature of action and its relation to our motivating reasons and intentions, they find themselves at odds with neuroscientific accounts that provide causal explanations for how neural signals are processed in the brain, resulting in specific bodily movements. And when philosophers confront ethical questions, they grapple with reconciling moral intuitions and principles with evolutionary biology and psychology, which offer accounts of how moral behaviours and judgements have developed through natural selection and social conditioning.

While this tension appears profound, it also exposes philosophy to ridicule, particularly in a society deeply entrenched in scientific thinking. Consider, for instance, the longstanding debate sparked by the simple statement, 'I see a tomato.' This statement has ignited extensive discussions about whether it expresses the act of *seeing a tomato* or merely *seeing the surface of a tomato*.[7] To many, such debates may seem absurd, as most people would expect an explanation detailing how light reflects off the tomato's surface, strikes our retinas, and is subsequently processed by the brain.

In response to debates that operate entirely independently of the sciences, the science-minded philosopher Daniel Dennett has dismissed many philosophical discussions as superfluous:

> There are quite a few philosophers who do very well what they do, executing their self-assigned tasks with cleverness, technical proficiency, and rigor, but engaged in an enterprise that's just not worth doing. (Dennett, 2013, p. 97)

2.4 The Clash in the Material World

We need not delve into discussions of free will or normativity to observe a clash between the sciences and our everyday manifest concepts and experiences; this tension is evident even when we consider the most mundane objects in our daily lives, such as chairs, tables, mountains, apples, houses, and stones. These are what we refer to as *ordinary objects*, or *medium-sized dry goods*.[8] Unlike abstract objects such as numbers, or purely fictional objects like Aladdin's magic lamp, ordinary objects exist within time and space. Ordinary objects are distinct from theoretical

[6] The Libet experiment, conducted by neuroscientist Benjamin Libet in the 1980s, demonstrated that brain activity associated with making a decision occurs milliseconds before the individual becomes consciously aware of their intention to act. This finding suggests that unconscious processes may initiate actions before conscious awareness (Libet et al., 1983), and it has been used to argue against the existence of free will (see, e.g., Rosenberg, 2011). Even though I believe that free will and determinism are compatible, the debate and challenge result from the tension between scientific discovery and ordinary, manifest experiences.

[7] The debate originates in the work of Clarke (1965).

[8] The term 'medium-sized dry goods' was preferably used by the philosopher J. L. Austin (1962, ch. 2).

entities like elementary particles, which are postulated in the sciences or discovered using experimental devices. They are simply the material items we encounter in our everyday lives.[9]

As we look around, we perceive a wealth of ordinary objects, each possessing shared or unique properties. We might take a gold ring in our hand, feel its smooth surface glide effortlessly along our palm, and observe its continuous form and shiny, yellowish hue. Yet we know that science tells an entirely different story about this piece of metal. Between 1908 and 1913, Hans Geiger and Ernest Marsden, under the direction of Ernest Rutherford, conducted experiments in which a beam of helium nuclei was fired at thin metal foil. They discovered that most of the helium nuclei passed unimpeded through the metal, with only a few being deflected. This finding radically challenged our intuitive understanding of the world and the prevailing physical theories of the nineteenth century.

At the time, the most popular theory of atomic structure was the 'plum pudding model,' which suggested that matter consisted of a positively charged, continuous 'pudding' filled with electrons. Rutherford's experiments shattered this notion, replacing it with a radically different model, according to which the material universe is largely an empty void. The gold ring in our hand is not the continuous structure it appears to be but rather a discontinuous collection of atoms with vast empty spaces between them.

In light of these scientific discoveries, it is with good reason that the physicist Sir Arthur S. Eddington famously suggested that our scientific descriptions of material objects stand in stark contrast to our ordinary, common-sense perceptions of those same objects. Eddington illustrates this point in his discussion about the nature of tables:

> Yes; there are duplicates of every object about me – two tables, two chairs, two pens. . . . One of them has been familiar to me from earliest years. It is a commonplace object of that environment which I call the world . . . It has extension; it is comparatively permanent; it is coloured; above all it is *substantial*. . . . Table No. 2 is my scientific table . . . My scientific table is mostly emptiness. Sparsely scattered in that emptiness are numerous electric charges rushing about with great speed; but their combined bulk amounts to less than a billionth of the bulk of the table itself. (1928, ix–x)

While the first table corresponds to the familiar, easily understood object of common sense, there also exists the non-substantive scientific table, consisting of atoms but mostly empty space. The two images appear irreconcilable with each other, both on ontological as well as on explanatory grounds. While, for instance, a table as a manifest object is homogeneous and continuous, the scientific theoretical postulation of a table as being constituted of atoms is non-homogeneous and non-continuous.

---

[9] Some academics distinguish ordinary objects from living beings. This distinction is only relevant in the present book when mentioned explicitly; otherwise, plants, animals, and humans can also be considered ordinary objects.

Could it not be that manifest objects are simply large groupings of individual particles, possessing different properties from those of their constituent parts? Not according to Sellars, who writes that '... the principle of reducibility [which pertains to the scientific image] ... makes impossible the view that groups of particles can have properties which are not "reducible to" the properties and relations of the members of the group' (p. 35). For him, the manifest table is not reducible to a set of particles, and thus the manifest object is not identifiable with the scientific object. It is, rather, an appearance to the human mind. In other words, given our cognitive faculties and biological constitution, particles are arranged in ways that appear to have certain properties from a human perspective. We are physical creatures, typically around one to two metres in size, and on this basis we can only perceive the atomic world as having a continuous form. This way of seeing the world is in this sense only an appearance of how the world actually is.

Summary

In this chapter, I identified the general problem of the clash between the manifest and the scientific image, which gives rise to the question of whether the objects of our experience are real. I presented a specific understanding of a metaphysical project that seeks to distinguish existing from non-existing objects, using Quine's notion of existence. We can now proceed to analyse the outcome of this enterprise with regard to ordinary objects. As we shall see in later chapters, this project ultimately fails by treating existence as a univocal category—one that purports to carve reality at its joints—when, in fact, this assumption proves untenable.

Chapter 3
Eliminating Ordinary Objects

3.1 Seeking a Simple World

Given Quine's notion of existence, different metaphysical views have emerged regarding the ontological status of ordinary objects, in other words, understandings of how ordinary objects fit into the general structure of reality. As a starting point, most of these ontological endeavours treat ordinary objects as composite objects. A table, for instance, is composed of different parts, such as four legs and a surface. These parts are themselves composed of smaller wooden parts, which are ultimately composed of fundamental *particles*. By definition, these particles are no longer further decomposable.[1] The ontological question about ordinary objects is then translated and generalised into the question of whether composite objects exist.

One popular response is to deny the existence of composite objects. This position, often referred to as *eliminativism*, is rooted in Quine's naturalist outlook and reflects an emphasis on the value of a sparse ontology. The principle of parsimony states that, in theory formation, entities should not be introduced beyond necessity. It is typically attributed to the scholastic philosopher and theologian William of Ockham and is known as *Ockham's razor*, though its roots can be traced back to Aristotle.[2] By this principle, we ought to prefer an explanation of unfamiliar patterns in the sky that appeals to peculiar bird migrations over one that invokes UFOs, even if both explanations are equally consistent with the empirical data. Unnecessary entities ought to be eliminated, and when two competing theories explain the same phenomena, the simpler or more parsimonious theory should be preferred. Given

[1] Note that 'particles' are used here to refer to any object that physics postulates to be most fundamental. These can also be strings, waves, or other entities yet to be discovered. To use a term that explicitly avoids commitments to current physical theories, many metaphysicians prefer to use the term 'simples' instead of 'particles'.

[2] See Charlesworth (1956) for a discussion of the origins of Ockham's razor and its appearance in Aristotle's *Physics*.

A. C. Schwaninger, *Cognitive Metaphysics*, Synthese Library 529,
https://doi.org/10.1007/978-3-032-05420-3_3

that the Quinean approach to metaphysics treats the sciences as the ultimate point of departure for answering existence questions and since the sciences themselves rely on the principle of parsimony, some have argued that '[t]he Quinean method is eliminativist by design' (Schaffer, 2009, p. 372). That is, for many metaphysicians in the Quinean tradition, we have good reason to eliminate superfluous entities from our ontology.

But even if we agree with Quine that the sciences provide the best candidate objects to write into 'the book of the world', consisting of a list of all existing things, why should we employ the principle of parsimony and focus only on the entities postulated by physics, rather than also including composite objects as they occur in other sciences (e.g., carbohydrate molecules in chemistry, trees in botany, or humans in psychology)? While it might be important for the sciences to formulate theories that are not more complex than necessary, metaphysics is not required to endorse such constraints for pragmatic reasons. Nevertheless, many metaphysicians regard parsimony not merely as a virtue that enhances comprehensibility or aesthetic appeal, but as something that enables us to uncover the true structure of reality. As Ted Sider states: '[S]implicity is a guide to truth, together with the thought that eliminating primitive notions makes a theory "structurally" simpler' (Sider, 2013, p. 240). Although some argue that eliminativism remains ultimately driven by pragmatic considerations, and that metaphysical theories, in this regard, are not so different from scientific ones (Benovsky, 2016), there remains a clear worry that a non-parsimonious theory might license the inclusion of Zeus or an undetectable teapot orbiting the sun in our ontology.

In accordance with these considerations, the eliminativist rejects composite objects. That is, he claims that objects composed of other parts should be eliminated from our ontology.[3] Since the fundamental particles of physics constitute all material objects in the universe, and since these particles, together with the laws of motion, fully describe all occurrences in the material world, there is no need to introduce any further objects into our ontology, such as trees and tables (Benovsky, 2018; Merricks, 2001).

Trenton Merricks, who is among the most prominent proponents of this view, argues that his position is the most compatible with our scientific understanding of the world, regardless of what physical theories precisely state.[4] He argues for his position as follows (Merricks, 2001). Let us assume there are N atoms in a room arranged *tablewise*. In that case, there are N objects in the room. However, if we assume that tables actually exist, there would not be N but rather $N + 1$ objects in the room, where one of these objects is *identical* to N others by reduction. This position, so Merricks argues, is unacceptable for various reasons. First, the position violates the principle of parsimony, as the table, being an additional object to the

[3] For a detailed discussion, see Korman (2015).

[4] This is not entirely accurate, since, just like van Inwagen, Merricks makes an exception for living systems and claims that these are the only composite objects that exist. In what follows, I disregard this exception unless explicitly stated otherwise.

other N objects in the room, is causally redundant, meaning that the table itself has no additional effect on the rest of the world if it existed. When plates and dishes are lying on the table, the table as a composite object is not supporting them in any additional way beyond what the other N objects already do. Thus, to assume that the table exists is to assume that a causally redundant entity exists. Second, the position is said to be implausible because a one-to-one identity relationship[5] cannot hold, as it entails mereological essentialism—the view that a composite object has its parts necessarily—which is typically rejected by straightforward analysis: if one particle were removed from a table, then the table would cease to exist. To assume the existence of ordinary objects is to assume composition as identity. Given that the latter assumption is false, there are no ordinary objects from this scientific perspective.[6]

If there are only particles and no composite objects, how then should we think about, and refer to, the objects within the manifest image? Instead of stating that there are chairs or tables, most eliminativists encourage us to say that there are only 'particles arranged chairwise' or 'particles arranged tablewise'.[7] By introducing these phrases, they provide a sense in which our ordinary parlance about everyday objects remains true, even if those objects do not exist.[8]

Particles arranged tablewise appear to compose an object, namely a table. Because of this, one might wonder whether we do not already have empirical evidence from the manifest image that there really are tables and chairs. Reliance on our experiences is, however, not an option for the eliminativist. As most metaphysicians believe, existence questions, like most metaphysical questions, cannot be resolved on the basis of empirical investigation, since the existence of composite objects has no effect on the causal structure of the world, including our senses. Although it looks as if ordinary objects exist, the eliminativist, like most other metaphysicians, would say that the world would look exactly the same whether or not tables and chairs existed, because all the causal work that allows me to see a table through my visual system is done by the particles of which tables and chairs are composed.

[5] Merricks follows the relational definitions of Lewis (1991).

[6] Note that eliminativism treats particles as indivisible, smallest physical units (Schaffer, 2003; Sider, 1993; Zimmerman, 1996). However, it could in principle also be the case that no such smallest unit exists—an issue that might threaten the eliminativist position. Even though I do not subscribe to such a view, it could be that every physical entity is composed of still smaller parts, and that there is an infinite regress of composition. Considering the history of chemistry, this seems a plausible option: prior to the discovery of protons and neutrons, atoms were also thought to be the smallest units of matter.

[7] Some have expressed concerns about paraphrasing talk of ordinary objects using expressions like 'particles arranged tablewise' (McGrath, 2005; O'Leary-Hawthorne & Michael, 1996; Uzquiano, 2004). See Benovsky (2018, ch. 4) for an eliminativist account that claims to avoid such expressions.

[8] While for van Inwagen (1990) ordinary manifest judgements about ordinary objects are true despite their inexistence, Merricks judges such statements only as 'nearly' true (Merricks, 2001, p. 171).

3.2 Debunking Common-Sense Beliefs

For some, eliminativism is too extreme a position to take. On the one hand, it seems odd to simply dismiss the existence of all the material objects surrounding us. But for others, it appears even more mysterious that we would have to give up on our own existence:

> There is an obvious and apparently very grave difficulty attaching to the suggestion that [ordinary objects] do not exist. This is that we ourselves, or at least our bodies, seem to qualify as such objects. And we certainly shouldn't be comfortable with the suggestion that we ourselves don't exist. Nor should we be comfortable with the suggestion that, although we ourselves exist, our bodies do not. (Lowe, 2005, p. 512)

While we might not be able to empirically determine the existence of tables, we are surely able to determine our own existence, or so the argument in Cartesian spirit goes. But this concern is not very persuasive, since the eliminativist will simply reply that living beings are composed of particles that *jointly* think they exist and *jointly* have the inner experience of a unified self. Again, in the causal structure of the universe, it makes no difference whether I exist—not even for my own inner experiences—or whether there are only particles arranged in a particular way. Thus, on these grounds, there is no good argument to think that the particles form a single thing (Rosen & Dorr, 2002).[9]

A perhaps more convincing argument against eliminativism interprets Ockham's razor differently. If we say that entities should not be multiplied beyond necessity, and by this mean *logical necessity*, then we arrive at the undesirable view of solipsism (i.e., that only we exist), which is entirely at odds with our ordinary understanding of reality. In contrast, if we interpret the principle as stating that entities ought not to be multiplied *without good reason*, then it also cautions against reducing our ontology without justification (Maudlin, 2007, p. 3). If there is no good reason to eliminate a part of our ordinary ontology, then it is worth preserving, even in light of Ockham's razor.

Does this reply save our common-sense ontology and avoid the clash altogether? Can we simply say that protons, electrons, and Higgs bosons exist, while at the same time chairs, tables, and people also exist? Not so fast. There are still many reasons why eliminativism remains attractive, some of which are provided by so-called *debunking arguments*. This term refers to a specific argumentative strategy that dismisses a manifest belief due to its lack of explanatory connection to facts within the scientific image. These types of arguments are not only employed in the context of material objects, but are most commonly found in metaethics and aesthetics, where they serve as arguments against the existence of universal and objective ethical or aesthetic principles.

---

[9] Note that van Inwagen and Merricks, who both fiercely defend the view that almost all composite objects do not exist, make an exception for living beings. Thus, they are only eliminativists with regard to non-living composite objects.

A debunking argument proceeds by offering a scientific explanation of how manifest beliefs are formed, without presupposing the truth of those beliefs. For instance, the type of moral realism which holds that there exist moral facts, in a Quinean understanding of metaphysics, can be challenged through a two-step debunking strategy. First, one can trivially observe that no physical theory postulates any entities that could serve as moral entities. This, of course, is not sufficient to undermine the realist position, as the realist might simply respond that physical theories are incomplete. Thus, the second step is to provide an aetiology, that is, an explanation of the causal origins of our moral beliefs that does not rely on the existence of physical moral entities. In this case, an aetiological explanation can be drawn from the Darwinian theory of natural selection. According to this theory, natural selection has shaped us to survive as a species, not to track truths about morality, even if such truths exist. In order to sustain themselves as a group, homo sapiens evolved into empathetic creatures who suffer when faced with unjustified violent acts toward others. This empathy increases group stability and improves the chances of survival, since each individual gains certain benefits by taking care of others. It would therefore be a mere coincidence if our responses to certain occurrences in the world happened to track universal moral facts that are not even identified by our best physical theories.[10]

When it comes to material objects, it might at first be less clear how debunking arguments could apply, given that our ordinary beliefs about the structure of reality are clearly shaped by our senses, which are the basis of all empirical knowledge. However, upon closer examination, debunking arguments are said to be just as successful in this context. In van Inwagen's *Material Beings* (1990), where he introduces the so-called *special composition question*, preliminary considerations are laid out for the first step in debunking composite objects. Given two material objects *A* and *B*, the special composition question asks under what circumstances their physical composition, *AB*, exists. Van Inwagen considers several initially plausible-sounding candidate answers to the question. One such candidate is *contact*. That is, one might suggest that any contact between *A* and *B* results in the new composite object *AB*. Drawing on ordinary experience, van Inwagen concludes that if contact were the correct criterion of composition, this would lead to highly implausible consequences: whenever two people shake hands, they would briefly constitute a single object. Contact seems far too unstable a relation to serve as the basis for composition. Another plausible-seeming candidate answer is that two objects, *A* and *B*, compose something when they are in contact by *force* and can only be separated by applying a stronger force that breaks the composite object *AB* into separate parts. Yet this too fails to provide a satisfying result. Imagine two people who, while shaking hands, become paralysed and are unable to separate. Would the moment they become paralysed mark the creation of a new composite object

[10] For an overview and critical examination of various evolutionary debunking arguments in ethics, see Shafer-Landau (2012). A critical analysis of the success of such metaethical arguments is provided in FitzPatrick (2015).

made of both people? Paralysis adds nothing to the furniture of the world, and so van Inwagen dismisses this option as well. Other possibilities he examines, such as gluing objects together or having them be physically bonded, also fail to yield an objective criterion that explains how certain objects form a unity while others do not. Although some objects appear to constitute a composite entity, van Inwagen argues that every attempt to identify a physical or ontological basis for their unity ultimately fails. Any such criterion that aims to explain what makes several parts form a whole turns out to be either arbitrary or inadequate.

Apparently, compositionality is yet another aspect of the manifest world that stands in tension with the scientific image. Based on van Inwagen's analysis, the prospect of finding a universal answer to the special composition question that is grounded in objective physical principles and aligns with our ordinary manifest beliefs about cases in which objects appear to compose is slim. Although the particles of which ordinary objects are constituted are said to exist, the unifying substance or 'thing' that holds them together to form a single entity does not seem to be part of the objective world, as every proposed objective criterion appears to yield cases that violate our experience-based intuitions.

Compositionality, as we ordinarily experience it, cannot be traced back to the physical domain. Van Inwagen does not, however, provide any aetiological explanation of when composition occurs, but he maintains that his analysis requires no further justification (van Inwagen, 1990, p. 36). He therefore presupposes that an aetiological explanation needed to complete the debunking argument could, at least in principle, be provided—if not today, then in the future, once the relevant sciences have matured sufficiently.

Other philosophers have also developed arguments suggesting that our experiences cannot be causally reduced to mere physical phenomena. If this is the case, then it is no longer necessary for the eliminativist to provide all the aetiological details in order to debunk our ordinary beliefs about tables and chairs. One of these arguments is the *argument from arbitrariness*. Imagine a situation in which there are particles arranged treewise located right next to particles arranged dogwise. We could of course say that in this situation we have a dog and a tree next to each other. But we could just as well say that there is a *trog*, defined as a trunk and a dog, along with some other objects. Or we might say that there is an *uptree*, defined as particles arranged treewise standing upright, which ceases to exist when turned over.[11] Ontologically speaking, it appears to be arbitrary which objects we choose to refer to. Imagine you are holding a pen in your hand. You might say that the pen is one object and that everything you are holding constitutes that object. Let us assume that the pen has a lid on it. What are you holding in your hand when you remove the lid? Is it now a different object? Or do the pen and the lid still form one object despite no longer being in contact? Depending on the momentary

[11] These are examples from Korman (2014).

Fig. 3.1 There appears to be no ontologically unifying principle governing the particles in (**a**), since a variety of different grouping options such as (**b**), (**c**), and (**d**) exist. None of these options seems objectively better than the others

circumstances and our practical needs, the way we choose to refer to certain things varies. Generally speaking, when we observe particles as illustrated in Fig. 3.1a, we might immediately have a sense of how they are grouped together, as indicated in Fig. 3.1b. However, as shown in Fig. 3.1c and d, we might just as well decide to group the particles in very different ways, depending on our needs.

The arbitrariness by which compositionality could occur suggests that compositionality is not an aspect of the world itself but is instead determined by factors such as our social practices in speaking about objects or the biological contingencies that shape how we perceive them. From an objective point of view, there appears to be a multiplicity of possible ways in which particles could compose. Moreover, there is no objective explanatory connection between our ordinary beliefs about a table and the scientific conception of a table. Given both the vast number of possible ways in which particles could compose and the absence of any explanatory connection between our ordinary beliefs and the physical world, we must conclude that our beliefs are shaped by how we have evolved as a species and how we have adapted to a particular language community. Since neither our biology nor our linguistic conventions are necessarily aligned with the structure of the world itself, our ordinary beliefs can only be true by pure coincidence. As Mark Heller writes:

> We conceptually divide up the world into objects one way rather than another because doing so will serve our purposes better . . . there is little chance that the resulting ontology will be the true ontology. . . . In principle, we could by sheer coincidence arrive at the true ontology. . . . I will discount the possibility of such a coincidence. (Heller, 1990, p. 44)

As in the case of debunking arguments, this line of reasoning highlights the absence of objective, unifying principles at the physical level and assumes that no further entities are needed to explain why unification seems to occur in one way rather than another within the manifest image. Both arguments take the scientific image as given and remain unconcerned with how our manifest experiences arise, as this is regarded as irrelevant to metaphysical inquiry.

3.3 More Puzzles Concerning Ordinary Objects

As we have seen, believing in our common-sense ontology leads to difficulties, since there appears to be no satisfying ontological answer to the special composition question other than either eliminating all ordinary objects or accepting that any arbitrary collection of a sufficient number of particles forms an ordinary object. The problems we encountered above result directly from adopting a scientific perspective and comparing it to our manifest worldview. However, many puzzles concerning ordinary objects can be traced back to ancient philosophy and do not depend on a scientific understanding of the world. Rather, the following puzzles about ordinary objects arise directly from manifest considerations. They are puzzles and paradoxes that provide further reason to abandon our ordinary, common-sense ontology, since they would never arise if we did not treat ordinary objects as existing entities.

The first of these puzzles concerns the question of identity and whether an object that has all its parts replaced is still the same object. The puzzle is famously illustrated by the ship of Theseus thought experiment, which can be stated as follows: Theseus sails with his ship around the world. At each harbour he visits, one of the ship's planks is replaced with a new one. By the end of his years-long journey, all planks have been replaced. Did Theseus end his journey aboard the same ship with which he had begun? And if we imagine that someone had collected all the replaced planks and reconstructed the ship in exactly the same form as it had originally been, would this ship be the same as Theseus's original ship, or as the final mended ship, or as neither?

The puzzle can also be formulated as a paradox, as presented by Thomas Hobbes:

> Two bodies existing both at once would be one and the same numerical body. For if, for example, that ship of Theseus, concerning the difference whereof made by continued reparation in taking out the old planks and putting in new, the sophisters of Athens were wont to dispute, were after all the planks were changed, the same numerical ship it was at the beginning; and if some man had kept the old planks as they were taken out, and by putting them afterwards together in the same order, had again made a ship of them, this, without doubt, had also been the same numerical ship with that which was at the beginning; and so there would have been two ships numerically the same, which is absurd. (Hobbes, 1655, p. 11)

The paradox appears to result from two conflicting intuitions about ordinary objects. On the one hand, we tend to think that objects preserve their identity through a continuous causal history, with parts being added or subtracted over time, and that they cease to exist once this causal history is disrupted by disassembly. On the other hand, we also tend to think that two objects are identical if they are composed of exactly the same material parts (Hoffman & Rosenkrantz, 1997, ch. 5.4).

Each intuition on its own seems to suggest that Theseus's ship does not simply cease to exist during the journey. However, it appears to be a completely arbitrary choice to conclude that Theseus's ship is either the mended ship or the reconstructed ship. Moreover, it seems equally wrong to say that Theseus's ship is both the mended and the reconstructed ship, since this would imply that the two ships are identical, that is, that they share all properties. For obvious reasons, this cannot be true, as the ships are located in different places and are made of different planks. For the eliminativist, the only viable option is to deny the existence of ships altogether (Korman, 2011; van Inwagen, 1990, pp. 128–135).[12]

The second puzzle concerns the non-collocation of ordinary objects.[13] Imagine a sculptor who possesses a lump of clay that we shall call 'Lump'. One day, the sculptor forms a statue from Lump that portrays the biblical figure of King David. Before he begins crafting, David, the statue, does not exist, whereas Lump does. After his work, the statue appears to exist. Clearly, the statue is not identical with the lump of clay, since the two have different modal properties, that is, they have different survival conditions. The lump of clay existed long before the statue came into existence and would continue to exist even if it were flattened by a truck, whereas the statue would at that point cease to exist. But if Lump and the statue are not identical, this implies that two distinct objects are located at one and the same place, which we ordinarily find implausible.

While some have argued that we ought to accept this seemingly problematic conclusion (Lewis, 1976; Wiggins, 1968), or that Lump ceases to exist during the period in which David exists (Burke, 1992), the eliminativist's strategy is once again straightforward: eliminate statues and lumps altogether (Unger, 1979; van Inwagen, 1990). All that really exists are particles, which do not concern themselves with our common-sense ways of naming and tracking objects in the world, and which cannot be assigned different modal properties.[14] By eliminating objects such as statues,

[12] This thought experiment also illustrates how our intuitions about the ontological status of material objects shift depending on the kind of object under consideration. For instance, most of the cells in a human body are replaced over a period of about seven years. Yet as long as there is continuity in the subject's mental states, we tend to think that the person remains the same. In the case of living beings, we are thus more inclined to believe that only the continuous object is numerically identical with the original. In contrast, when we consider historically significant artefacts, such as the Mayflower, we are more likely to believe that only the reconstructed version is truly the Mayflower, and not the mended one. See Hoffman and Rosenkrantz (1997, ch. 5.4), who argue on this basis against the existence of artefacts.

[13] Notice that the thought experiment of Theseus' ship can also be said to illustrate issues of collocating objects. On the one hand, one could think that at the end of Theseus' journey, his ship is located in two distinct places at the same time. On the other hand, one could also think that two different ships were initially collocated at the beginning of Theseus' journey.

[14] Notice that both the problem of coinciding objects and the previous puzzle concerning identity rest on Leibniz's Law, according to which two objects are identical if and only if they have all the same properties. Some have suggested that these puzzles may be resolved by rejecting this law (Geach, 1967).

there is no longer any threat that two composite objects could coincide, because they simply do not exist. The puzzles are thereby resolved in a rather brutal manner.[15]

Puzzles three and four arise from examining what happens when we begin altering the boundaries of an object. In what follows, I shall speak in terms of particles, although the puzzles are just as troubling if we simply replace 'particle' with '1mm^3 of matter'. As long as all matter can be divided into small cubes, the puzzles remain. They do not require any insight or reflection concerning the structures that may lie beyond these tiny cubes at smaller scales. Nevertheless, for the sake of convenience, I shall formulate the puzzles in terms of particles.

Puzzle number three runs as follows. Let us imagine a room containing N particles arranged *dogwise*. According to common sense, these particles constitute a dog. Let us call this particular configuration of particles 'Lassie'. Now consider any subset of these N particles from which a single particle has been removed, resulting in a set of $N - 1$ particles. This new subset clearly also forms a dog, since the removal of a single particle does not mark the difference between being a dog and not being a dog. If Lassie is the dog composed of N particles, then any configuration consisting of $N - 1$ particles must also be a dog. The question now is whether the dog composed of $N - 1$ particles is still Lassie, or whether it is a different dog.

If we assume that $N - 1$ particles constitute a different dog, then there is not only Lassie in the room but also at least one other dog. In fact, we should expect that there are extremely many other dogs present. Since a dog consists of an unimaginably large number of particles (on the order of magnitude of 10^{27}, if we are speaking of atoms), and we assume that each subset of $N - 1$ particles constitutes a different dog, then there must also be an unimaginably large number of dogs located in the room. This appears to be problematic, as we would normally expect there to be only one dog in the room. This third puzzle is often referred to as the *problem of the many*, and has been cited as evidence that ordinary objects cannot exist, since we would otherwise be led from intuitive assumptions to unacceptable conclusions (Unger, 1980). For the eliminativist, it is clear that the choice between whether the N or the $N - 1$ particles (or both) are said to be Lassie is entirely arbitrary, since Lassie does not exist in the first place.

A fourth puzzle arises if we instead answer affirmatively to the question of whether $N - 1$ particles still constitute Lassie. In this case, we are faced with the problem of *metaphysical vagueness*. If $N - 1$ particles also constitute Lassie, and if the removal of a single particle makes no difference to whether the object is Lassie, then, logically speaking, even a single particle must be Lassie. This is clearly absurd. Once again, for the eliminativist, this result is no more troubling than the original presupposition that ordinary objects exist at all, and the puzzle is resolved by eliminating such entities altogether.

[15] For a concise overview of existing metaphysical positions on coinciding objects, see Wasserman (2009).

3.4 Is Eliminativism the Holy Grail of Metaphysics?

While there are many metaphysical positions that provide alternatives to our common-sense ontology and are not discussed here,[16] we have seen that there is a strong case to be made in favour of eliminativism, which became the theory of choice for many during the late twentieth and early twenty-first century. Eliminativism aligns closely with Sellars' distinction between the scientific and manifest images. It treats the scientific image as the correct point of orientation for understanding reality as it truly is, and dismisses the manifest image as an arbitrary construct shaped by biological and social contingencies. Not only does eliminativism appear attractive from a scientific standpoint by offering a parsimonious ontology grounded in physical theory, it also benefits from other virtues that resonate with the scientific spirit, for example, providing an ontology that is easier to work with and to explain (Tallant, 2013).

Most importantly, eliminativism is, for many, a superior option in light of the concerns raised by ordinary objects. It appears to offer a robust worldview that withstands a wide range of philosophical challenges and supplies straightforward answers to the puzzles discussed above. As Jiri Benovsky puts it, he favours this position because:

> Eliminativism is an incredibly useful tool for solving metaphysical problems. In short, eliminativism works, it nicely solves or avoids many problems that alternative views have—and that's the best reason to endorse it. (Benovsky, 2018, p. 6)

Does this mean that we have now solved all metaphysical questions about material objects? To achieve a complete understanding of reality, do we merely have to wait for physics to determine which particles are the ultimate constituents of the world, while the task of the metaphysician is effectively complete? As we shall see, alternative metaphilosophical approaches, some of which were prominent in earlier periods of intellectual history, have gained renewed momentum in recent years and cast doubt on the supposed finality of eliminativism.

Summary

Eliminativism is a natural choice within the Quinean approach to metaphysics, as it aligns closely with the perspective offered by contemporary physics. Numerous puzzles concerning ordinary objects appear to undermine our common-sense understanding of the world. Debunking arguments lend considerable support to the eliminativist position by highlighting a lack of explanatory connection between our beliefs about ordinary objects and the underlying physical facts. That is, our beliefs about the existence and identity of ordinary objects do not seem to be causally or

[16] While eliminativists reject all composition, a popular alternative is universalism, which takes the opposite view, namely, that any composition forms a new object, including, for example, the object composed of the nose of Michelangelo's David and the Eiffel Tower (Lewis, 1986a, 1991). Many of the puzzles surrounding ordinary objects can also be addressed using this approach. For further discussion, see Korman (2015, ch. 3.1).

epistemically grounded in the scientific image of the world. Due to the seemingly arbitrary nature of how the world is divided into distinct ordinary objects, the eliminativist concludes that such objects simply do not exist. This conclusion is further supported by the observation that a host of perplexing puzzles arises once we assume the existence of composite objects like trees and tables, but not of countless equally arbitrary configurations such as togs.

Chapter 4
Eliminating Intuitive Metaphysics

Eliminativism offers a bold and seemingly final response to the tension between the manifest and scientific images: if ordinary objects cannot be reconciled with fundamental physics, then perhaps they do not exist at all. By denying the reality of such objects, eliminativism claims to dissolve the puzzles that arise when we try to fit tables, chairs, and trees into a world of particles and wave functions. What remains, it suggests, is no longer a metaphysical problem but a scientific one: physics will tell us what exists, and metaphysics, having done its part, may quietly retire.

Yet this vision of metaphysical closure has not endured unchallenged. In recent years, doubts have emerged not just about eliminativism itself, but about the entire framework of thinking in which it gained prominence. Increasingly, philosophers have begun to question whether the terms of the debate have been poorly set from the start. These deeper reflections are known as *metametaphysical* concerns: they target not this or that ontological position, but the very assumptions, aims, and methods that have shaped modern metaphysical inquiry.[1]

These concerns are no less threatening to eliminativism than to the views it opposes. For they challenge the broader metaphysical project in which eliminativism is embedded—one that assumes existence questions are central, that physics sets the ontological standard, and that the task of metaphysics is to investigate how a science-oriented ontology relates to the objects of our experience by formulating and resolving various puzzles involving both manifest and scientific entities.

In this chapter, I argue that the debate from which eliminativism emerged is fundamentally flawed, insofar as it rests on an intuition-driven methodology. This

[1] The term 'metametaphysics' implies the existence of a standard approach to metaphysics—most prominently the science-oriented, Quinean tradition—which, like many of its historical predecessors, is once again under scrutiny. Since one's metaphysical method is itself guided by substantive assumptions about reality, it is not clear that the line between metaphysics and metametaphysics is more than a disciplinary artefact.

A. C. Schwaninger, *Cognitive Metaphysics*, Synthese Library 529,
https://doi.org/10.1007/978-3-032-05420-3_4

does not undermine eliminativism—in fact, it strengthens the case for it. But it does mean that eliminativism must be understood more radically than many of its advocates have supposed. Only in the next two chapters will it become clear that the Quinean approach, with its emphasis on existence questions, is misguided in a more fundamental sense. Other questions, I argue, must take precedence if metaphysics is to engage meaningfully with the structure of reality.

4.1 The Use of Intuitions in Metaphysics

So far, we have examined some of the arguments metaphysicians invoke to motivate a sparse ontology. Many of the considerations intended to support the eliminativist position regarding ordinary objects derive from some form of *intuition*. On the dialectical level, when a speaker puts forward an intuitive claim, they are typically not expected to supply further reasons for it. While such claims may be supported by inferential justification, they are generally taken to require no additional defence, as they simply *seem* to be true. Intuitive claims have therefore been characterised as non-inferential dialectical justifications, and in this sense, have been said to function as the premises of philosophical arguments (Chalmers, 2014).

The appeal to intuitions is both appropriate and, in many cases, indispensable across various domains of inquiry. For instance, whenever the outcome of an experiment or an argument is assessed, it is typically presupposed in an intuitive way that our memory of the initial conditions or premises has not failed in the meantime and is generally reliable. It would therefore be mistaken to criticise the use of intuitions in general. Nevertheless, an excessive reliance on intuitions is troubling for several reasons: intuitively plausible propositions may still be false, they may conflict with the intuitions of others, and they often lack independent justification. Although many philosophers have defended the use of intuitions as a valuable tool in philosophical theorising, others have expressed deep reservations, particularly in the context of metaphysics.[2]

Let us now consider the role of intuitions in some of the arguments encountered so far. Recall the debate about compositionality, which concerns the question of when two material objects together form a new, composite object. According to the common-sense, conservative view, composition sometimes occurs, whereas eliminativism holds that it never occurs. Yet there are also other positions. *Universalism*, for example, maintains that composition *always* occurs, such that even my nose and the Eiffel Tower jointly compose an object of *strange kind*.[3] Although

[2] For a recent defence of intuitions in philosophy, see Climenhaga (2018). A broad critique is developed in Cappelen (2012) and Williamson (2007), while a focused attack on the role of intuitions in metaphysical theorising is found in Ladyman and Ross (2007).

[3] Only very few philosophers hold the view that composition occurs exclusively in accordance with our experiences. See, for example, Markosian (1998).

both extreme views, universalism and eliminativism, clearly diverge from ordinary intuitions, part of the reasoning that supports one over the other typically involves an intuitive assumption. Eliminativism, in particular, is often defended on the basis of an intuitive, evidence-based premise that objects of strange kind do not exist:[4]

P_1: Objects of ordinary kind exist if and only if objects of strange kind exist.
P_2: Objects of strange kind do not exist. [intuitive premise]
C: Therefore, eliminativism is true.

Premise P_1 is motivated by the view that it would be irrational to affirm the existence of objects of ordinary kind while denying the existence of equally arbitrary, strange composites. If we were to do so, it would seem almost miraculous that reality coincides so neatly with our ordinary way of carving it. Interestingly, P_1 is accepted by the universalist as well. However, in contrast to the eliminativist, the universalist argues that perception provides evidence for the existence of ordinary objects and therefore rejects P_2, replacing it with the claim that objects of ordinary kind do exist. Given P_1, it then follows that composition must occur indiscriminately, and so universalism is true.

One may also recall the argument presented against eliminativism that draws upon the intuition of one's own existence. It can be summarised as follows:

P_1: I know that I exist. [intuitive premise]
P_2: If I exist, I have proper parts.
C: Therefore, there exists at least one composite object.

The eliminativist is willing to accept P_2 but, as we have seen, will reject P_1 in favour of claiming that 'particles jointly believe that they exist', a formulation from which the conclusion C does not follow. While the non-eliminativist may attempt to offer further reasons for endorsing P_1, these arguments ultimately rest on an intuition about one's own significance in the world.

Further arguments clearly grounded in intuition include van Inwagen's examples involving two people shaking hands, which he uses to support his conviction that there exists no uniform criterion for when composition occurs. Although he finds it 'pretty clear that one cannot bring a composite object into existence by bonding two human beings ... to each other' (van Inwagen, 1990, p. 62), he himself concedes that such claims are ultimately intuitive. In the case of two objects merely touching, he writes:

> It is a basic conviction of mine that this theory [that two things compose a third at contact] is wrong and that its being wrong is in no sense a matter of convention. I cannot prove this thesis, for I know of no propositions more plausible than itself from which it could be derived. I can only say that I shall try to display in this book the fruits of agreeing with me about this and various similar theses. (van Inwagen, 1990, p. 36)

[4] This example is analysed in detail by Higgins (2014, ch. 3).

Given that van Inwagen's conviction rests on intuition, it is unsurprising that alternative views have been developed which, while relying on the same scientific facts, propose different conclusions.[5]

Most sides within a metaphysical debate about ordinary objects appeal to intuitions in support of their respective positions. However, despite this reliance on intuition in the formulation of arguments, the resulting theories often radically contradict intuitions. In the case of compositionality, both the universalist and the eliminativist appeal to intuitions they consider plausible from a common-sense perspective, yet they ultimately conclude that our common-sense ontology is entirely mistaken. As alternatives to the ordinary view, they recommend either removing or adding an enormous number of objects from or to our ontology. While intuitive assumptions are frequently expected and often taken for granted in metaphysics, it is nonetheless considered acceptable for the resulting theories to diverge significantly from common-sense.

For readers familiar with other areas of metaphysical inquiry, it will be apparent that the reliance on intuition is not confined to debates about ordinary objects. As surveyed in detail by Benovsky (2015), intuitions also play a prominent role in metaphysical discussions of time. This includes the debate over whether time is a dimension akin to space, in which objects are simply located (B-theory), or whether time genuinely passes, thus generating an ontological distinction between present, past, and future (A-theory). Some philosophers argue that the strongest motivation for accepting the A-theory stems from the nature of our experience and the intuitions it generates. As Bradford Skow puts it: 'I cannot survey all the motivations philosophers have had for the moving spotlight theory. But the motivation that I like best appeals to the nature of our conscious experience' (Skow, 2009, §4). This illustrates how our experience shapes intuitions that influence metaphysical theorising, even if they do not constitute decisive arguments. Indeed, many philosophers endorse the B-theory despite its apparent conflict with ordinary intuitions.

4.2 Critique from Scientifically Minded Philosophers

The use of intuitions is not inherently problematic; on the contrary, it is a necessary feature of any theoretical enterprise, including the natural sciences. However, the excessive reliance on intuition raises significant concerns. As we have seen, metaphysics does not suffer from the problem that theories must conform to our intuitions—eliminativism and universalism clearly do not. What, then, is the precise concern voiced by those who criticise metaphysics for its use of intuitions?

[5] See, for example, Husmann and Näger (2018) who think that bonding does determine compositionality.

The main objection is not that metaphysicians reject propositions merely because they are unintuitive. Rather, the concern is that the foundational concepts employed in metaphysical inquiry are typically drawn from everyday language, folk psychology, or outdated scientific theories, such as high-school physics, rather than from contemporary scientific practice. For example, metaphysicians often rely on notions such as compositionality, simples, causation, and objects. These concepts are not merely different from those used in fundamental physics; they are, in many cases, *incompatible* with the conceptual framework of modern physics. This incompatibility risks producing a distorted or misleading picture of how the subatomic realm relates to the macroscopic world.

While metaphysicians speak of objects as being composed of simples or particles, modern physics suggests that this intuitive way of conceptualising material objects may be deeply problematic. As Ladyman and Ross observe, contemporary metaphysics

> seeks to account for the world as 'made of' myriad 'little things' in roughly the way that (some) walls are made of bricks. . . . [However, according to contemporary physics] it is no longer helpful to conceive of either the world, or particular systems of the world that we study in partial isolation, as 'made of' anything at all. (Ladyman & Ross, 2007, p. 17)

For many philosophers with a background in physics, the idea that the macroscopic world is fundamentally built from small building blocks appears implausible, if not misguided. From the perspective of quantum mechanics, it is even questionable whether it makes sense to speak of separable individuals in the first place (Esfeld, 1999a). Ladyman and Ross go so far as to ridicule micro-reductionist arguments, pointing out that such arguments presuppose a conception of reality that finds no support in modern quantum theory, and instead reflects an understanding of physics 'learned in A-level chemistry' (Ladyman & Ross, 2007, p. 24).

The physics we learn in school is largely classical Newtonian mechanics. Although Newtonian mechanics was once considered highly unintuitive—especially due to Newton's notion of force—it is nonetheless a theory that explains and predicts phenomena largely in accordance with our everyday experiences. Newton's theory of matter aligns well with Democritus's atomism, which remains the basic intuition about matter for many non-physicists today. If physics had stopped at this point, we might not need to provide any account of ordinary objects beyond the perspectives offered by Merricks and others. However, the situation has been radically transformed by the advent of quantum mechanics, at least if we suppose that the features peculiar to quantum mechanics not only govern the microphysical domain but also extend to the macroscopic world. Under this assumption, we are faced with deeply perplexing problems when attempting to apply quantum-theoretic concepts to material objects of ordinary size. This is famously illustrated by Schrödinger's cat thought experiment, which shows that if we apply the principles of quantum mechanics to the macroscopic level, we must consider a cat in a sealed box to be both dead and alive simultaneously, until we look into the box (Schrödinger, 1935). According to quantum mechanics, quantum systems can be in a state of *superposition*, meaning that they occupy multiple states at once. If this is also true of macroscopic systems, then a cat, which is presumed to be just a very large quantum system, must likewise be capable of occupying multiple incompatible states, such as being both alive and dead. Which state the system ends up in is only determined when a measurement is made, that is, when we open the box and observe the cat. This thought experiment is, to say the least, profoundly counterintuitive and far removed from common-sense reasoning. It reveals just how problematic the relationship is between the quantum and the macroscopic domains. Expressions from metaphysics that appear to bridge these two worlds, such as 'particles arranged catwise', do little to clarify the true nature of reality.

The fallacies committed by metaphysicians when discussing ordinary objects and claiming to uncover deep principles about reality become even more apparent in the context of quantum entanglement, particularly in relation to the issues of localisation, separation, and identification. The phenomenon of entangled particles refers to the correlation of their fundamental properties in a way that cannot be attributed to mere chance. If one adopts a realist interpretation of quantum mechanics, then quantum mechanics and classical mechanics cannot be reconciled by appealing to hidden variables. In that case, quantum entanglement must be regarded as an objective feature of nature, existing independently of any observer. On this basis, quantum entanglement challenges three central assumptions commonly made about ordinary objects (Esfeld, 1999a,1999b). First, contrary to our classical intuitions, quantum systems and their properties are generally not *localised* in specific regions of space-time. Second, entangled systems cannot be described independently of one another. From the standpoint of quantum mechanics, this undermines the classical principle of *spatial separability*, according to which the state of each individual system determines its local properties (Howard, 1989). By contrast, the local properties of entangled systems appear to be determined by the joint state of the system as a whole. Third, entangled particles cannot be considered distinct *individuals*, since there exists no property by which one can be distinguished from another (French & Redhead, 1988). If the properties of entangled systems in quantum mechanics are correctly interpreted as not supervening on their parts, and if quantum entanglement extends beyond the microscopic to the macroscopic domain, some have concluded that the whole is ontologically prior to its parts (Esfeld, 1999a; Maudlin, 1998). On the further assumption that all matter interacted at the time of the Big Bang, it has been argued that all particles may in fact be entangled, forming a single universal system (Gribbin, 1984, p. 229). Although it remains uncertain whether entanglement is indeed universal, it appears implausible—on physical grounds—to assume that the ordinary objects of our experience can be clearly separated and distinguished from one another in any fundamental ontological sense.

For the metaphysically well-informed reader, another example of a scientifically problematic position in metaphysics is David Lewis's theory of Humean supervenience. This is a case in which a philosopher explicitly acknowledges that his view depends on empirical assumptions, yet appears to rely primarily on what one might learn during secondary education. Lewis's theory has been criticised for standing in stark contradiction to the findings of modern physics, as it ultimately rests on an ontology of points localised in space-time, with perfectly natural, intrinsic properties instantiated at those points. This picture conflicts with key features of quantum mechanics, in particular the phenomena of non-locality and entanglement—concepts that were already well established among physicists at the time Lewis formulated his theory (Humphreys, 2013). Although Lewis openly admitted that his theory involved 'a contingent, therefore an empirical, issue' (Lewis, 1986a, p. xi), it seems that he did not fully take into account, or at least failed to address, the significant challenges that contemporary physics poses to his metaphysical framework. This may reflect a lack of familiarity with the relevant science, or an underestimation of its implications for foundational metaphysics.

4.3 Metaphysics Is Not Prior to the Science

In Aristotle's original conception, metaphysics was understood as 'first philosophy',[6] and the discipline has long held the reputation of addressing questions that are prior to any scientific inquiry. While some domains of metaphysics may still retain this character—though often at the cost of ceasing to provide an ontology relevant to our reality—much of contemporary metaphysics no longer operates independently of science. In fact, much of what metaphysicians do today is anything but prior to the sciences. This point is easily illustrated by considering the absurdity of a metaphysician building a theory based on pre-Newtonian physics. As Ladyman and Ross note: 'If it really doesn't matter that classical physics is false then we might as well do our metaphysical theorising on the basis of Aristotelian or Cartesian physics. But then the absurdity would be patent' (2007, p. 26). Consider, for instance, Descartes's claim that 'the virtue or power of self-movement found in one body may indeed pass wholly or partially into another and thus be no longer present in the first; but it cannot entirely cease to exist in the world' (Descartes, 1988, p. 85). One can easily imagine a metaphysician informed by Cartesian physics constructing a detailed account of bodily virtues. However, no modern metaphysician would engage in such an endeavour, not because Newtonian mechanics is more aligned to our pre-theoretical physical intuitions than Cartesian physics, but because metaphysicians, like anyone else educated in our society, operate within the conceptual framework of science they have learned.

Newtonian mechanics is, in many ways, counterintuitive. It is only through a thorough education that one acquires a sense of how its concepts work and learns to employ them productively. Metaphysicians, like the rest of us, have become so familiar with the Newtonian and atomistic worldview that its presuppositions often go unnoticed. To take just one example, we have so far assumed that ordinary objects are composed of simples. As discussed earlier, this is neither a trivial nor a purely non-empirical assumption. Physics challenges the notion that ordinary objects are

[6] He uses this term most prominently in *Metaphysics* E.1.

made up of simples, and we must turn to empirical investigation to determine whether such entities exist. It is just as much an empirical claim that there are any simples at all. Indeed, the physicist and Nobel Laureate Richard Feynman gives this assumption such weight that he writes:

> If, in some cataclysm, all of scientific knowledge were to be destroyed, and only one sentence passed on to the next generations of creatures, what statement would contain the most information in the fewest words? I believe it is the *atomic hypothesis* (or the atomic fact, or whatever you wish to call it) that all things are made of atoms—little particles that move around in perpetual motion, attracting each other when they are a little distance apart, but repelling upon being squeezed into one another. (Feynman, 1963, p. 4)

If metaphysics is not supposed to be based on any empirical findings then we could not even make the assumption that ordinary objects are objects composed of particles, since it is logically just as possible that a non-discrete substance is holding all things together. Whether this is truly the case is for physicists to decide and metaphysicians must acknowledge that the concepts they develop are in a continuation of these findings.

Whether or not we should conceive of reality as composed of simples or whether this does not do justice to the findings of quantum mechanics is not the question here. Instead, the point is twofold: first, our manifest concepts alone are largely inadequate to provide a general theory of reality and second, metaphysics has and always will be dependent on some empirical findings if one tries to account for an actual rather than merely possible realities.

There is nothing inherently problematic about metaphysicians constructing their theories in light of scientific facts. The problem arises, however, when they fail to recognise that their claims depend on scientific presuppositions. Metaphysicians such as Merricks make assertions about the relationship between ordinary objects and their underlying constituents, yet often assume that all they need to know about the empirical world is that ordinary objects are composed of some kind of substance. They refer to these fundamental building blocks as 'particles' or 'simples', and believe that, from the empirical side, their philosophical work can proceed entirely from the armchair. For this reason, many metaphysicians falsely assert that their theories are compatible with *any* fundamental physical theory, and are therefore true regardless of whether the ultimate physical description of the world refers to strings, bosons, or something else entirely. Furthermore, when metaphysicians fail to acknowledge that their claims hinge on substantive scientific assumptions, they are unlikely to engage seriously with the relevant sciences and are thus prone to employ inadequate or outdated conceptual tools in their theories.

4.4 Armchair Metaphysics and the Clash

One might wonder why the sciences are more likely than philosophy to develop better concepts for characterising fundamental reality. The reason lies in the empirical nature of scientific investigation, which involves active experimentation rather than passive reflection. Through the use of experimental devices, scientists interact with the world in ways that give rise to concepts which could never have been conceived through armchair contemplation alone. Consider, for instance, the double-slit experiment: it reveals that light exhibits a dual nature, behaving both like a particle and a wave depending on the specific configuration of the experimental setup. It is difficult to imagine anyone arriving at the concept of wave-particle duality purely through reflection on their everyday experiences. Accordingly, it is the sciences that must guide metaphysics in selecting the initial concepts from which its theorising may proceed.

This is where we again encounter the *clash*. Our theorising may be shaped either from within the armchair, by reflection on immediate experience, or by empirical investigation, which can uncover features of the world that require entirely new conceptual tools. These empirically grounded concepts are often unintuitive and could not have been predicted in advance. When philosophers construct new concepts, they do so using the intellectual resources available to them in the armchair. While such methods may be fruitful in certain areas, as will be discussed later in the book, they prove inadequate in the context of metaphysics.

Summary

Metaphysics faces two central problems. First, it is still commonly conceived as a practice that is *prior* to the sciences, an assumption that, at least in the case of material object metaphysics, is mistaken. Second, many of the scientific claims upon which metaphysical theories depend are either inaccurate or fundamentally misguided, often relying on concepts that physicists would never employ. A fundamental physicist, for example, would not think to use the notion of 'composite objects' in formulating a theory of the world. Composition is the wrong concept for making the right distinctions when attempting to carve reality at its joints. If metaphysics aims to reveal the nature of reality in its most general form, without becoming entangled in empty reflections on conceptual schemes that contribute nothing to our understanding of the world we inhabit, it cannot avoid integrating scientific concepts into its theoretical frameworks. Any account of what it means to be an individual that ignores the insights of contemporary science falls short of what good metaphysics ought to achieve.

Chapter 5
Metaphysics Is Not (only) About Existence

In the previous chapter, I discussed the problem of conducting philosophy from the armchair and how such an approach can result in theories that are simply inadequate in light of contemporary scientific developments. This observation, however, does not undermine the core convictions of eliminativism. On the contrary, it may even strengthen them. The eliminativist might now be inclined to say: 'I should never have been misled into debating such misguided philosophical problems, which attempt to understand the world by starting from concepts rooted in the manifest image. I will simply reject ordinary objects, compositionality, and all the other spurious notions philosophers have invented, and instead commit myself solely to the concepts provided by physics: entanglement, superposition, the wave function, and so forth'. A central motivation behind eliminativism lies in its ability to avoid troubling philosophical puzzles. If these puzzles arise from relying on flawed and scientifically unsupported concepts; so much the better. The eliminativist can argue that it is not merely the result of socio-cultural and biological contingencies that certain collections of matter appear to form new objects, but that the very notion of compositionality is metaphysically misguided and can be debunked on scientific grounds. The concept of compositionality does not assist us in carving mind-independent reality at its joints.

Consider, for instance, a philosopher such as Tim Maudlin, who possesses a deep understanding of fundamental physics. In his work, he does not engage with the debates discussed in the previous chapters—debates that arise from puzzling circumstances within the manifest image. Instead, he focuses exclusively on questions concerning how we ought to interpret quantum mechanics:

> metaphysics, insofar as it is concerned with the natural world, can do no better than to reflect on physics. Physical theories provide us with the best handle we have on what there is, and the philosopher's proper task is the interpretation and elucidation of those theories. In particular, when choosing the fundamental posits of one's ontology, one must look to scientific practice rather than to philosophical prejudice. (Maudlin, 2007, p. 1).

A. C. Schwaninger, *Cognitive Metaphysics*, Synthese Library 529,
https://doi.org/10.1007/978-3-032-05420-3_5

For Maudlin, metaphysics must be informed by the sciences and can support their endeavour to discover what there is. He has no interest in the traditional paradoxes and can be regarded as a truly *radical* eliminativist with respect to manifest objects, since he sees no need or value in engaging with debates that rely on manifest concepts. These concepts, being rooted in intuition, often lead only to unnecessary difficulties. While I believe Maudlin's research is of great value, I contend that it rests on too narrow an understanding of metaphysics, as the following line of thought will suggest.

5.1 The Deflationary Perspective on Metaphysics

The way I have described the metaphysical enterprise so far reflects a metaphysical realist position, in which eliminativism is one of many possible positions, only one of which can ultimately be true. Traditionally, the metaphysical realist maintains that we have epistemic access to a unique, mind-independent world, and that truths about the structure of this world can be discovered. This position, however, is not without its critics. In a thought experiment, Hilary Putnam (1987) imagines a world consisting of only three individuals x_1, x_2, and x_3, and asks us to consider how many objects there are in that world. For the eliminativist, the answer will be three. For others, the answer will be the sum of all possible combinations in which the three basic individuals can compose, which is seven: $x_1, x_2, x_3, x_1+x_2, x_1+x_3, x_2+x_3$, and $x_1 + x_2 + x_3$. For yet others, the answer will be eight, since one must add to the previous list a 'null object' that is included in every object. According to the metaphysical realist, there exists a unitary and objective sense of what it means to be an 'object', and thus the question 'how many objects does the world contain?' is perfectly meaningful and has a determinate answer, even if we remain uncertain about what it is.

For the realist, there exists one description of the world that is privileged over all others, and it is this privileged description that Putnam (1981) seeks to challenge. In the absence of a privileged linguistic scheme, it is, according to Putnam, absurd to suppose that there is one objectively correct answer to the question of which of the above descriptions of his imagined world of three individuals ought to be preferred. He argues that, for any sentence to possess meaning, we require a conceptual scheme—or, more loosely speaking, a linguistic framework—that determines the meaning of the terms involved. Whenever we attempt to answer a question such as how many objects there are, we do so within the boundaries of the 'object' concept as defined by our linguistic framework, and our answer will accordingly reflect that framework. However, if no such framework is specified, and since there exists no universally accepted framework, the question itself lacks a determinate answer. General questions about the existence of certain objects cannot be answered in the abstract, for they are ill-formed in the absence of an appropriate linguistic context.

This aspect of Putnam's account closely aligns with Rudolf Carnap's rejection of traditional metaphysical questions. Carnap is often portrayed as the intellectual antipode to his student Quine, particularly regarding the possibility and legitimacy of metaphysics. In much of the traditional metaphysical literature, their disagreement is typically framed as having a clear victor: Quine. However, more recent scholarship suggests that Quine's position is not as distinct from Carnap's as many Quinean metaphysicians have assumed, and Carnap's philosophy may not have been conclusively defeated, as is often claimed.[1]

Carnap's rejection of metaphysical questions is grounded in his notion of a linguistic framework and his distinction between 'internal' and 'external' questions (Carnap, 1950). He defines a linguistic framework as a set of terms and associated rules that determine how those terms are to be used in discourse. Internal questions are then questions that can be answered within a given framework by logical, analytical, or empirical means. For example, in school mathematics, we learn to operate within a particular linguistic framework for discussing numbers. If we are asked in an exam how many prime numbers there are between 10 and 15, we can provide a logically verifiable answer: 2 (namely, 11 and 13). Or during a night excursion with a physics teacher, students might be asked whether there is more than one moon orbiting Jupiter, and the correct answer can be empirically verified by pointing a telescope at the relevant celestial bodies. In both cases, the question concerns the existence of something from within an established linguistic framework, and in both cases the existence question can be answered by anybody familiar with the framework. Likewise, for anybody familiar with the framework, it would seem entirely inappropriate to respond to such questions with statements like 'there are no numbers and therefore also no prime numbers' or 'ordinary objects like Jupiter and moons do not exist'. This is because the linguistic framework in use provides clear standards and procedures for answering the teacher's questions.

In the case where someone asks the general questions 'Are there numbers?' or 'Do ordinary objects exist?', Carnap believes that the opposite situation applies. He calls these types of questions *external questions*. These are questions that concern 'the existence or reality of the framework itself' (Carnap, 1950, pp. 21–22). For Carnap, external questions lie outside any framework and are therefore ill-defined, lack cognitive content, and have no uniform answer. He expresses this concern, for instance, when considering the metaphysical debate on whether numbers exist:

> Unfortunately, these philosophers have so far not given a formulation of their question in terms of the common scientific language. Therefore our judgment must be that they have not succeeded in giving to the external question and to the possible answers any cognitive content. Unless and until they supply a clear cognitive interpretation, we are justified in our suspicion that their question is a pseudo-question, that is, one disguised in the form of a theoretical question while in fact it is non-theoretical. (Carnap, 1950, p. 209)

[1] For a detailed discussion of how Quine differentiates himself from Quinean philosophy, see Egerton (2020). For a general overview of the Carnapian programme, see Kraut (2020). In what follows, I will primarily rely on Price's (2009) and Thomasson's (2015) interpretations of Carnap's deflationism.

Ontologists raise highly general questions that are never raised by 'the man on the street' and that turn out to be pseudo-questions if asked with the expectation of a factual answer. Asking an internal question *uses* a term appropriately within the context of a framework, whereas asking an external question merely *mentions* the term, with the appropriate usage still to be determined (Price, 2009). Rather than understanding external existence questions as questions about what really exists, they can be seen as practical questions concerning which linguistic framework we ought to adopt. External questions may prompt us to develop a framework of our own, specifying how the mentioned terms are to be appropriately used. For instance, it may sometimes be useful to adopt an eliminativist framework, allowing us to discuss reality from a purely physical perspective, whereas in other contexts, such as when grocery shopping, a common-sense ontological framework may be more appropriate: when asking the store manager where the milk is, we presuppose its existence simply by operating within the relevant framework.

Carnap's approach to metaphysics is what is known as *deflationism*. This is the general view that many philosophical debates and concepts, particularly in metaphysics, are less deep or substantial than they initially appear. According to this perspective, disputes over existence, truth, or reality often stem from confusion about language or from the lack of a clearly specified conceptual framework. Rather than seeking to uncover a single, objective structure of the world, the deflationist maintains that metaphysical questions should be approached in terms of the pragmatic utility of different frameworks. Questions that are clear and uncontroversial within a given linguistic framework can become ill-formed once removed from that specific context. Deflationism thus treats many metaphysical questions as either pseudo-problems or practical matters of conceptual choice, rather than as attempts to reveal the true structure of reality.

To give another example of the deflationist position, consider the metaphysical debate about free will. For the metaphysical realist, this debate has a right or wrong answer. For the deflationist, we are free to choose a linguistic context that suits our purposes. In a physics laboratory where particles are discussed as following determinate trajectories, the notion of free will may be difficult to accommodate, making its rejection useful. By contrast, in a courtroom, it is vital for the judge to distinguish between individuals who are free to choose their actions and those who are not (e.g., due to external coercion or limited mental capacity). As Carnap puts it: 'we have to make the choice whether or not to accept and use the forms of expression in the framework in question' (p. 23). This decision requires philosophical reflection and an evaluation of the trade-offs a given framework entails. Consequently, eliminativism is not to be dismissed; it remains a viable framework for certain purposes, particularly when aiming to characterise reality in a parsimonious, scientific way. However, according to the deflationist, it is merely one of many conceptual options and therefore does not invalidate the adoption of other ontological frameworks.

Those readers who harbour a certain resentment towards metaphysics due to its obscure questions and never-ending discussions that appear to lead nowhere might find Carnap's insights appealing, for they explain why metaphysicians do not converge in their views: it is simply because they are engaged in different linguistic enterprises. Other readers, however, might be disappointed, for they have conceived of ontology as the quest to unlock the deep, fundamental secrets of reality. They might have hoped that metaphysics could offer the curious, intellectual mind a way to unravel the true nature of the world and to understand what binds its innermost core together through pure reflection. Instead, the discipline turns out to be nothing more than a pragmatic evaluation of which linguistic framework serves our purposes best. Nevertheless, although existence questions may lose their initial appeal, our broader interest in metaphysics need not fade. We will later discover new questions for metaphysics to pursue. As we have seen earlier, the Quinean approach identifies metaphysics with ontology—the study of what exists—and this has so far shaped the way we have analysed the clash: by demarcating existing from non-existing entities. However, the deflationist shows that questions of existence are either pseudo-questions or questions that can be resolved with ease, at least from the philosophical side.

5.2 Easy Ontology

For Carnap, the deflationist viewpoint amounts to a rejection of metaphysics. For someone like Amie Thomasson, however, deflationism does not entirely abandon metaphysical work but reduces it to linguistic problems that, while still worth engaging with, are comparatively easy to resolve. Her so-called 'easy' approach to ontology provides answers to ontological questions by analysing how language is used in different contexts. As she argues, existence claims can only be meaningfully evaluated once we have specified a domain of quantification. This means that the objects within the domain must be enumerable and have clear conditions of existence for any judgement about their existence to make sense (Thomasson, 2007, ch. 6.3). The technical philosophical term often used for such enumerable objects is *sortals*. When Putnam asks how many 'objects' there are in his imaginary world and claims that the answer depends on the chosen linguistic framework, he draws attention to the fact that 'objects' is not a sortal term (Hilpinen, 1996). If we were to substitute 'objects' with a sortal term like 'alligators' and ask how many alligators there are in a given world, a unique answer could be given through observation (Thomasson, 2007, p. 120).[2]

For Thomasson, the differences in opinion regarding whether or not a thing X exists concern *only* variations in how we can understand X, *not* in how we can interpret the notion of existence itself. This is important, as it consolidates her realist stance. Putnam, by contrast, appears to believe that the notion of existence may change from one conceptual scheme to another. Many deflationists have approached

[2] For further discussion on whether and how sortal terms apply to countable objects, see Wiggins (2001) and Lowe (1989, 1997).

existence in this way, namely that the existential quantifier ∃ can vary between frameworks. According to Eli Hirsch (2002), the quantifier has different meanings for different philosophers, such that disagreeing ontologists are often merely talking past one another.[3]

Taking the existential quantifier to be invariant implies that when an eliminativist talks about the existence of Higgs bosons, the qualification is the same as when one says that tables and chairs exist. Claiming to have established a realist interpretation of the Carnapian enterprise, Thomasson believes that she can thereby resolve the clash between the manifest and scientific images. According to her, there simply is no clash, because our scientific and ordinary discourses operate with distinct and disconnected ontologies. Eddington's two tables can only be said to clash if they are referring to the same thing. According to Thomasson, however, reference to objects is determined by the framework in use. If the frameworks differ, so too do the things being referred to, and thus there is no clash:

> scientific theories ... do not use sortals such as 'table', and if science and common sense are using sortals of different categories, the 'things' picked out by the two descriptions cannot be identical. (Thomasson, 2007, p. 142)

5.3 The Anti-realist Dimension of Deflationism

Carnap's ideas directly influenced Putnam's work and thinking. However, the key difference between the two for the present discussion is that Putnam interprets the subjective dimension of Carnap's position as having a metaphysical dimension. Since the deflationary approach allows the subject to decide which ontology to adopt for its current purposes, Putnam draws conclusions that make it appear as if reality is simply a construct of the subject or the mind. According to Thomasson (2015, ch. 1), this 'unfortunate turn' towards anti-realism is one of the main reasons why Carnap's deflationist approach was largely dismissed by subsequent philosophers.

Although Putnam, in this phase of his philosophical career, calls himself an 'internal realist' and explicitly tries to distinguish himself from the idealist camp by stating that '[h]uman minds did not create the stars or the mountains' (Putnam, 1990, p. 30), some of the metaphors he resorts to suggest that he does toy with the idea that reality is to some extent a mental construct. For instance, he makes claims that indicate strong anti-realist tendencies, such as '"objects" are as much made as discovered' (1981, p. 54), or 'If one must use metaphorical language, then let the metaphor be this: the mind and the world jointly make up the mind and the world' (Putnam, 1981, p. xi).

Humans' central role in constituting reality is, in Putnam's thinking, a consequence of our power to choose the linguistic framework we fancy: 'how many

[3] For a realist reply, see Sider (2004).

objects there are in the world ... is relative to the choice of a conceptual scheme' (Putnam, 1985, p. 32). However, Thomasson argues that the deflationist need not adopt any such anti-realist position in order to defend a Carnapian view of metaphysics. For her, we must simply make the right distinction between *using* a linguistic framework and being *dependent* on it.

It is evident that we must use some framework to talk about reality, but to say that reality itself depends on the framework is an entirely different story. While, according to Carnap, we form propositions and talk about facts by having a language framework, which includes noun terms and logical rules of usage, this view does not imply that facts are constructs of the subject or reducible to language. The rules of use for speaking about facts do not by themselves determine whether the correct application conditions are met. We may introduce noun terms like 'unicorn' or 'elephant', along with rules for speaking about material things, yet it is the world that dictates whether any of those terms have instances to which they apply. In this sense, the facts themselves are not subjective or mind-dependent.

5.4 An Evaluation of Deflationism

I agree with the deflationist idea that, in the absence of a specified linguistic framework, external questions are ill-formed pseudo-questions. Ontological inquiries must be assessed within such a framework to be meaningful. However, this does not mean that metaphysics is obsolete or reducible to a set of trivial problems. Rather, it indicates that metaphysics should not be identified with ontology. Instead, we should seek an alternative understanding of metaphysics that does not focus on existence questions, as these turn out to be philosophical trivialities at best.

For Thomasson, the clash is fundamentally a matter of linguistic interpretation. In her view, the physical sciences operate within a distinct linguistic framework, separate from the one we use in everyday life. This leads to a non-empirical strategy centred on analysing our conceptual system and how it shapes our thoughts about the world. But how convincing is this response to the *clash*? While a clear-cut distinction may be drawn at the level of language, different levels of description are not independent of each other. Treating the clash as a purely linguistic matter fails to acknowledge that reality forms a unified whole, and that tables and chairs are not isolated from the physical world, existing in a different sphere of reality. If we agree with Thomasson that ordinary objects are real things that exist in the world just like protons and electrons, while at the same time the physical entities also stand in some constitutive relationship to those ordinary objects, then we must also account for the structural differences that arise between these levels of reality.

Despite Thomasson's arguments against Putnam and her efforts to maintain a realist stance, characterising the clash as a linguistic issue and addressing ontology purely in semantic terms has prompted some philosophers to accuse her of departing from realism altogether. As Steven French (2019, p. 262) argues, distinctions in linguistic discourse can, in principle, be grounded in neuroscience, and any appeal to language or mental states could ultimately be replaced by scientific descriptions.

Even if we accept that existence claims must be made within a given context—an idea with which I am largely sympathetic—it does not follow that all domains of discourse contain noun terms that refer to objects of equal subjective import. Likewise, if the referents of terms are shaped, to some extent, by our cognitive structures, this does not entail that those objects fail to exist. Since existence claims are context-dependent, and since contexts may involve terms that vary in their subject-dependence, subjectivity and existence are distinct notions. This is evident in the fact that subjectivity admits of degrees, whereas existence does not. What matters, then, is how the structural distinctions between different levels of reality come about. In the next section, we will see that metaphysics demonstrates its greatest promise precisely in addressing this challenge and in developing ontological frameworks that integrate scientific insight.

Summary
By focusing on existence questions, metaphysics is misguided, as such questions can only be evaluated relative to a linguistic framework. However, conceptualising reality solely in terms of linguistic frameworks leads to an undesirable anti-realism that fails to offer a satisfying resolution to the clash. It is suggested that metaphysics redirect its focus away from existence questions and place greater emphasis on the clash itself.

Chapter 6
Levels of Reality: A Naturalist's Take on Metaphysics

In the previous chapter, we rejected eliminativism as a uniquely accurate account of reality. We saw that from a deflationary angle, existence claims need not be restricted to the entities posited by fundamental physics; objects in the manifest image, together with the entities recognised in chemistry, biology, psychology, and other sciences, may legitimately figure in our ontological bookkeeping, provided the relevant explanatory and linguistic frameworks are in place. Even if one is not persuaded by deflationism in full, for many philosophers the moral remains: physics does not exhaust the space of respectable ontological commitment. Once that door is opened, Sellars' model of the two images calls for re-evaluation so as to accommodate the diversity of scientific disciplines and the material objects to which they refer.

Motivated less by deflationism than by reflection on scientific practice, a number of philosophers—let us call them *liberal naturalists*—have questioned the supposed depth of the conflict between Sellars' two images. They argue that reality may be described at multiple length scales, and that the judgements made in physics or chemistry need not contradict those made at the macroscopic level; rather, they pertain to different descriptive levels, some of which may be more fundamental than others:

> Many philosophers now would acknowledge that the world may be described at many 'different levels'. This opens the possibility that characterizing the qualitative world we encounter in experience, the colours and shapes, the beach ball on the sand, and so on, is simply a matter of saying how things are 'at a different level' than the level of description used by the physicist. (Campbell & Cassam, 2014, p. 3)

It is certainly true that Sellars' binary distinction between a *scientific* and a *manifest* image is an oversimplification. The sciences do not form a unified enterprise, and different disciplines often present distinct and sometimes incompatible

A. C. Schwaninger, *Cognitive Metaphysics*, Synthese Library 529,
https://doi.org/10.1007/978-3-032-05420-3_6

perspectives on the world.[1] What we are left with, then, is a rich variety of *images*, typically emerging at different spatio-temporal scales or, more broadly, at distinct *levels of reality*. For the sake of simplicity, I will at times continue to refer to the twofold distinction between the manifest and the scientific image. However, much of the discussion to follow generalises naturally to a plurality of images.[2]

Ladyman and Ross (2013) have taken a leading role in establishing a liberal naturalist ontology: '[o]ntology is scale relative [which means that all empirically identifiable structures] except those of fundamental physics are detectable at some scales but not at others' (2013, p. 108). Under the label of 'scientific metaphysics', they promote a new movement aimed at radically naturalising metaphysics by dismissing the traditional armchair approach and adopting a more science-centred perspective. What characterises this movement is its refusal to appeal to intuition to motivate a priori claims, while—unlike the (logical) empiricists such as Hume or the Vienna Circle—retaining a positive metaphysical outlook. In contrast to Oppenheim and Putnam's (1958) foundationalism about physics, Ladyman and Ross argue that scientific metaphysics should not eliminate ordinary objects, but should instead aim to develop a *unified* account of the various independent scientific disciplines. Their motivation for preserving non-physical entities is not rooted in deflationist concerns, but rather in the fact that the sciences operate at different scales.

If we are to do justice to the sciences working at macroscopic levels, we should not eliminate the entities they posit. They propose that the task of unification should fall to metaphysics because this is what distinguishes it from the individual sciences. Each science is concerned with a specific domain of investigation, but no science is dedicated to producing a comprehensive, unified picture of reality. It is this gap that metaphysics is uniquely positioned to fill.

Ladyman and Ross accept the liberal, scale-relative approach to reality and seek unifying principles that offer an explanatory dimension beyond what the sciences in isolation can achieve. The unification they envision follows two main principles. First, they propose what they call the *Primacy of Physics Constraint*, which states:

> Special science hypotheses that conflict with fundamental physics, or such consensus as there is in fundamental physics, should be rejected for that reason alone. Fundamental physical hypotheses are not symmetrically hostage to the conclusions of the special sciences. (Ladyman and Ross, 2007, p. 44)

This ensures that the naturalist respects the authority of the empirical sciences and stands in opposition to dualist and emergentist positions. The second principle they endorse is the *Principle of Naturalistic Closure*:

[1] This point is elaborated in van Fraassen (1999), who challenges the rigidity of Sellars' dichotomy.

[2] It is even possible for tensions to arise between images situated at similar scales. As we shall see, levels of reality are often determined by the measuring instruments we employ, which tend to be scale-specific. Nevertheless, the core argument holds even in experimental contexts where fundamentally different methodologies apply at overlapping scales, potentially generating structural inconsistencies between co-located images.

> Any new metaphysical claim that is to be taken seriously at time *t* should be motivated by, and only by, the service it would perform, if true, in showing how two or more specific scientific hypotheses, at least one of which is drawn from fundamental physics, jointly explain more than the sum of what is explained by the two hypotheses taken separately. (Ladyman & Ross, 2007, p. 37)

From the standpoint of scientific unification, both principles are motivated by the fact that physics has the broadest explanatory scope of all the natural sciences. To achieve a sort of unification, this maximal scope must be preserved. The Principle of Naturalistic Closure allows for and even encourages metaphysical claims that build conceptual bridges between 'separately developed and justified pieces of science (at a given time) [that] can be fitted together to compose a unified world-view' (Ladyman & Ross, 2007, p. 45).

6.1 Making Sense of Ordinary Objects

The envisioned scientific unification for which Ladyman and Ross are hoping must also account for ordinary objects, as these are a central part of our empirical investigations and thoughts. As Peter Strawson writes: 'We think of the world as containing particular things some of which are independent of ourselves' (Strawson, 1959, p. 15). These things are central ingredients in our understanding of the world, and they cannot be eliminated if we still want to conceptualise the world we ordinarily encounter. As we have seen in the previous chapter, there may be a subjective element to the things we encounter in the manifest image, but this does not justify denying the existence of tables and chairs in favour of physical entities. Our phenomenological experiences of *things* must be accounted for in a naturalist metaphysics, not least because many of the special sciences formulate theories about classes of ordinary objects. Failing to account for our everyday experiences of a world populated by chairs, tables, and trees cannot be said to provide a unified picture of reality.[3]

This naturalist approach of unifying the sciences into a single overarching picture and reconciling it with the manifest image is not uncontroversial. Rosenberg argues that, throughout history, attempts at unification have consistently failed, which he takes to be a strong indication that the project is ultimately futile (Rosenberg, 2011, p. 218). Nonetheless, Ladyman and Ross propose a way in which we ought to conceptualise, in metaphysical terms, the unification between different levels of reality.

As we have seen, if we believe that quantum mechanics is approximately true and that the principles of localisation, separability, and individualisation lack any mind-independent physical basis, then any theory that posits ordinary objects

[3] Hofweber (2019a) even goes so far as to claim that we have empirical evidence for the existence of ordinary objects, and that some realms of science could not operate without them.

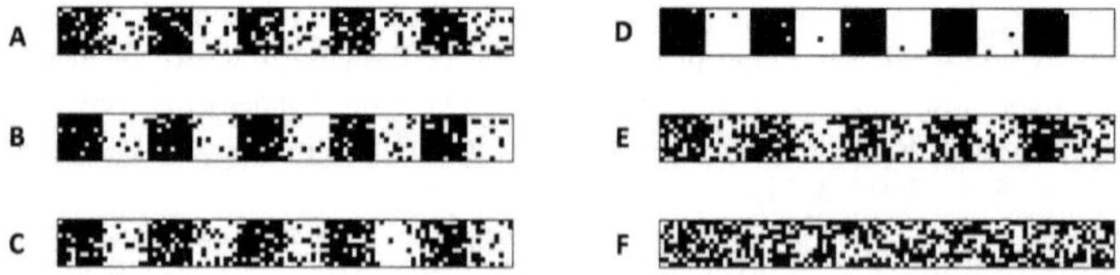

Fig. 6.1 Patterns in the data. Adapted from Dennett (1991)

in the classical sense will be in conflict with our most fundamental scientific understanding. But if ordinary objects are so central to how we engage with and understand the world, how can a unified picture of reality still be achieved in accordance with the Primacy of Physics Constraint? The answer provided by Ladyman and Ross is to treat ordinary objects not as individuals, but as *real patterns*—a term borrowed from Dennett (1991), which refers to relatively stable and enduring structural regularities within empirical data that allow for efficient description.[4] What does this mean?

Whenever we perform a scientific measurement, or simply observe the world with our eyes, we collect data. In computational terms, such data can be translated into a string of numerical values. A thermometer, for instance, might provide a sequence such as '23.2, 23.8, 23.7, 23.8', indicating temperatures measured at regular intervals. Another example is a digital image captured by a camera: the image is stored as a two-dimensional matrix of pixel values and can be flattened into a single string of numbers. Likewise, the information collected by our retina when we see an object can be interpreted as a stream of values that, if desired, can be represented in binary form such as '0110110100001001', much like how computers ultimately encode all data in terms of zeros and ones. Such a digital representation is often referred to as a *bitmap*.[5]

In Fig. 6.1, we see various two-dimensional visual representations of bitmap data, where zeros are rendered as white pixels and ones as black pixels. In the bitmaps A–E, there is a recognisable underlying structure: a regular alternation between 10×10 pixelated black and white squares. This repeating alternation constitutes a real pattern in the data—though in some cases (as in E), the pattern is partially obscured by noise, while in others (as in B and D), it is more readily visible. By contrast, bitmap F displays purely random data, in which no discernible pattern can be found.

A pattern is said to *exist* or be *real* if the data contains a non-random structure. While there is a real pattern in A–E, no such pattern is present in F. From an information-theoretic perspective, this means that real patterns are *compressible*, because structured data can be represented more efficiently than by simply listing

[4] The representation of any data is assumed to be efficient if its description requires less storage space than the full bitmap of the raw data.

[5] Strictly speaking, not all rational numbers can be exactly expressed in binary form. For instance, $\frac{1}{3}$ requires an infinite repeating binary expansion. In practice, however, such values are approximated to finite precision.

every individual bit. Thus, a real pattern can be expressed in fewer bits than the full bitmap. For example, the binary string '100010100111101' appears incompressible, as it must be recorded bit by bit. By contrast, '101101101101101' is clearly compressible: it can be represented more compactly as '5 × "101"' (five repetitions of the substring '101'). The principle behind real patterns is therefore not ontological in the traditional sense, but informational: patterns are real to the extent that they enable compact representation and predictive utility.

Importantly, compression is always relative to a given language or representational scheme. In the example above, the shortened description '5 × "101"' presupposes a language in which both the number 5 and the multiplication operator are expressible as atomic symbols. There is, therefore, no *absolute* measure of compressibility—only compressibility relative to a specified linguistic or computational framework.

For Dennett, patterns exist if they are *useful*—that is, if identifying them offers epistemic or practical advantages. In scientific modelling, such advantages arise from representational efficiency, since every model is, in effect, a form of compression. For instance, the best possible model of my hand, in terms of fidelity, is my actual hand. But this is not a model in any useful way, as it does not abstract or compress any information. A better model might be a simplified geometric representation—a sphere with five sticks attached. While clearly not identical to the hand itself, it omits for many usages irrelevant detail and thereby provides a compressed and tractable representation of the object.

Why are real patterns relevant to a scientific ontology? When we carry out empirical measurements, we look for structures in the data—structures that manifest through regularities or recurrent configurations. Such patterns must persist long enough to be registerable as a 'thing'. Purely random or unstructured data is of no scientific interest, as it can neither be detected nor modelled, and therefore cannot be said to be *real* in any meaningful empirical sense.[6]

According to Ladyman and Ross, what exists is what qualifies as a real pattern. This includes any structure that emerges from empirical data in a stable and compressible way. Importantly, this notion applies not only to medium-sized objects like chairs and tables, but also to what we ordinarily think of as the 'building blocks' of the universe, such as elementary particles. However, in their view, such particles are not fundamental 'individuals' in the traditional metaphysical sense, but rather enduring patterns within the mathematical structure of physical theory.

For Ladyman and Ross, the notion of an *individual*—especially as it appears in metaphysical discussions of ordinary objects—is deeply problematic. Unfortunately, they are not always precise about how this contrasts with real patterns. As Hawley notes, their conception of individuals seems to oscillate between several meanings, leading to interpretative difficulties: 'I have failed to work out whether or

[6] Note that some entities may be real by inference from real patterns. For example, the Big Bang—although a singular, temporally non-extended event—can still be said to be real, insofar as it is inferred from persistent cosmological patterns, such as redshift and background radiation.

where individuals (i.e. objects, I think) fit into the overall picture' (Stanford et al., 2010, p. 178).[7]

What is clear, however, is that Ladyman and Ross adopt an eliminativist stance towards individuals as discrete ontological units. Unlike traditional eliminativists who deny the existence of all manifest objects, they retain a role for objects as real patterns that emerge at different levels of reality. These patterns are not discrete, self-standing entities; rather, what we call 'objects' are 'epistemological book-keeping devices' (Ladyman & Ross, 2007, p. 240)—constructs shaped by our cognitive apparatus and pragmatic interests.

Instead of positing that ordinary objects exist as independent individuals, they propose to explain how the experience of such objects as *individuals* arises, without thereby conflicting with the scientific image. As Dennett puts it:

> The ontology of everyday life is now teeming with items that, like fatigues, sit rather awkwardly in the world of atoms and molecules. If we can understand how this population explosion *came about*, and why it is so valuable to us as agents in the world, we can perhaps discharge our philosophical obligations without ever answering the ultimate ontological question. (Dennett, 2013, p. 106, emphasis added)

For Ladyman and Ross, different real patterns emerge and exist at different scales. The table I am working on, for example, is a real pattern that exists at the scale of ordinary human perception—even if it is not, in fact, an individual, but only cognitively perceived as such. However, they argue that it makes no sense to say that the table exists at cosmic or quantum scales. This view ensures that the ontological commitments of the special sciences stand on a par with those of physics, and that there is no ontological superiority of physics over the other sciences. The only relevant asymmetry lies in the domain of *laws*: according to the Primacy of Physics Constraint, the laws of fundamental physics constrain those of the special sciences, but not vice versa.

This yields an important distinction: while the laws of physics enjoy a kind of explanatory priority, the *entities* posited by the special sciences are no less real than those posited by physics. Tables and chairs qualify as real patterns in precisely the same way that quantum fields or elementary particles do. They are not merely subjective or mind-dependent, and their description cannot be compressed further in any relevant information-theoretic sense.

6.2 Independent Levels of Reality

I agree with Ladyman and Ross that we should understand metaphysics as a unifying enterprise, yet, however promising their proposed metaphysical methodology may be, their ontology suffers from serious problems, as many have pointed out. Graham

---

[7] Hawley lists four possible interpretations of 'individuals' in Ladyman and Ross's work, but none provides a fully satisfying contrast with real patterns.

Harman (2010), for instance, critically examines their view and questions how a multiplicity of patterns and scales arises in the first place. He argues that the division into discrete scales and the existence of a plurality of real patterns is inconsistent with the notion of an objective, autonomous reality. Unlike other scientifically minded philosophers who eliminate all non-physical entities, Ladyman and Ross aim to avoid such a radical stance. But as a consequence, they must contend with the even more puzzling implication that their various levels of reality are simply disconnected from one another. As Harman puts it: 'the levels of Ladyman and Ross have no mutual influence at all. In their view, to say otherwise would merely amount to a folk poetry of cohesive individual things engaged in causal relations' (p. 779).

Ladyman and Ross treat the different layers as independent because they believe each level gives rise to emergent properties that are unexplainable and irreducible to the micro-level. The chaotic nature of particle behaviour at higher scales makes such behaviour unpredictable, and this justifies treating the levels as disconnected. The structures we perceive as persistent objects are not determined by a fixed set of underlying micro-level entities: 'there is no single candidate aggregate of real microscopic patterns that is best suited to be the reductive base of the everyday table' (Ladyman & Ross, 2007, p. 253).[8] On this basis, they reject Eddington's scientific table and take only the macroscopic table to be the real pattern that exists. The clash, on their view, is a natural consequence of the complexity of physical systems and the emergence of irreducible properties.

This response to the clash is no more successful than Amie Thomasson's suggestion that different scientific domains employ distinct linguistic frameworks. The problem with both views is that they fail to explain how the various levels and their descriptions are connected, and how they depend on each other, while at the same time appearing to be in a *clash*. Given that there is a sense in which reality is unified, a *scale-relative ontology* that postulates different entities at different levels—entities that are irreducible to one another and possess incompatible properties—faces a serious difficulty. This issue stems from an even deeper problem in Ladyman and Ross's ontology.

As Harman observes, their view contains a tension. On the one hand, it includes a realist commitment to a mind-independent world, which science helps us to uncover—particularly the view that reality is not composed of discrete individuals. On the other hand, they posit a plurality of patterns found at discrete levels of reality, which cannot be fully mind-independent, since compressibility is always relative to a representational system (potentially the brain's own language). Ladyman and Ross try to argue that the levels at which such patterns appear are guaranteed to exist independently of observers and then again it is unclear how the levels themselves can exist without an observer in someway determining at what scale a level starts

[8] John Horton Conway's *Game of Life* is a good illustration of how each level of reality 'has its own language' (Dennett, 1991, p. 39). At the lowest level, there are rules for when cells turn ON and OFF, whereas at a higher level we describe persistent, mobile formations using structural names.

and where it ends. Their attempt to merge an ontological claim about what there is with an epistemic claim about how we access truths about these entities falters. It falters because the world cannot be, at once, a unified whole with no individuals, and simultaneously a fragmented plurality of real patterns. As Harman writes:

> If the world itself were really just a monolithic lump, it is impossible to see why there would be myriad separate phenomena for an observer, especially since that observer should already have melted into the lump along with the rest of the cosmos. ... Any philosophy must find some way to account for both the quantized and the continuous aspects of the world, but not by programming their synthesis into the game in advance. For there is a paradox here, and work to be done. (Harman, 2010, pp. 787–788)

It may be that Ladyman and Ross's account would prove more successful if it incorporated an idealist dimension—one that acknowledges a subjective component as indispensable to any comprehensive metaphysical analysis. Without such a component, we struggle to make full sense of how the levels of reality come about. Yet, wary of being seen as advocating esoteric views, Ladyman and Ross repeatedly insist that their view bears no resemblance to Kant's idealism. To be sure, significant differences remain: for instance, unlike Ladyman and Ross, Kant never committed himself to the existence or non-existence of individuals in the mind-independent (noumenal) world.

Nevertheless, I believe that their central ambition—to construct a world-picture grounded in real patterns and to call upon metaphysics to unify the many domains we encounter—need not lead to paradox when complemented by a modest Kantian insight. If Ladyman and Ross were to accept that patterns only individuate and become a plurality of distinct entities through the perspective of an observer, then Harman's criticism would lose its force. The fragmentation into levels would no longer require full ontological independence, but could be seen as an effect of how cognition structures what is encountered. In this way, levels of reality remain real but are recognised as being accessible and distinct only from within a given perspective.

Summary
Liberal naturalism approaches metaphysics by adopting a science-oriented methodology and seeking to unify the various scientific domains. The positing of real patterns, observable at different scales, constitutes a metaphysical commitment within this view. Unlike eliminativism, which rejects all ordinary objects, the liberal naturalist affirms the reality of all levels. Yet this leads to a paradox: how can entities at each level be real while remaining independent of one another? A promising resolution may lie in acknowledging the subject as a necessary element in metaphysical theorising. Such a move has not yet been taken, despite the lack of any compelling naturalist reason to exclude it.

Chapter 7
Realism and Idealism

As we have seen, the objects that appear at different levels of reality do exist and must be accounted for in our metaphysical theory. Understanding the structure of reality requires building conceptual bridges between these levels, and much of metaphysics can be interpreted as the pursuit of a unified account of how they relate to one another. While Ladyman and Ross treat the levels as largely independent, any viable project of unification must explain how these levels arise and how they are connected. Treating them as entirely independent fails to acknowledge that they are all part of a single shared reality. We have seen that undertaking this task may require rethinking some of our basic realist assumptions. In the following chapter, we aim to clarify the debate between realism and anti-realism, in order to better understand what is at stake as we begin to develop an anti-realist or idealist account of reality in the chapters to come.

7.1 Idealism and Naturalism: A False Contrast

We have seen that Ladyman and Ross are at great pains to distance themselves from any suggestion that their position might be interpreted as idealist. Yet if incorporating an idealist element could help resolve some of the deeper problems in their theory, why are they so resistant to the idea? The answer, I believe, lies in a broader tendency among naturalist philosophers to reject anything that even vaguely resembles idealism. But is this widespread dismissal of idealism really justified?

So far, we have encountered a range of positions that fall under the label of *naturalism*. Most contemporary philosophers, including Quine, Ladyman, and Ross, identify themselves with this tradition. Broadly speaking, naturalism holds that there is nothing more to the world than what is disclosed by the scientific image. This claim has several dimensions. Metaphysically, it asserts that reality consists

A. C. Schwaninger, *Cognitive Metaphysics*, Synthese Library 529,
https://doi.org/10.1007/978-3-032-05420-3_7

solely of those entities to which the natural sciences are ontologically committed. Methodologically, it maintains that the methods and practices of the sciences are the only legitimate means of investigating reality.[1]

Most eliminativists are naturalists in this sense: they commit themselves to whatever entities fundamental physics ultimately postulates and regard the scientific approach as superior to our ordinary, intuitive ways of understanding the world. Although the liberal naturalist may not fully succeed in her ambitions, she too aims to align ontological commitment with the sciences. Unlike the eliminativist, however, she does so across multiple levels of description and sees value in relating these levels to one another.

Philosophers in the *idealist* tradition have often been portrayed as standing in sharp contrast to naturalism, as they prioritise the role of the mind over that of material objects. In some cases, such as Berkeley's idealism, this even entails the inclusion of God as a central ontological commitment. Idealist positions are typically associated with the view that 'the explanatory framework of the human subject is that which takes priority over other explanatory frameworks, including, crucially, that of nature and natural science' (Papazoglou, 2019, p. 99). As a result, idealism and naturalism are frequently regarded as mutually exclusive and could be one reason why Ladyman and Ross are eager to distance themselves from anything resembling idealism.

However, there is something curious about this contrast and the claim that naturalism and idealism are mutually exclusive. Both proponents of naturalism and even many of its critics seem to share the belief that the sciences have, in some sense, converged on a *realist* ontology that excludes the mind.[2] This assumption suggests that scientific inquiry has reached a stage where its core methods and frameworks are exhausted, and that no radically new image of the world is possible. It presupposes that the sciences have arrived at a unified worldview from which various philosophical disputes could, in principle, be resolved, one that privileges metaphysical realism over any view that places the mind at the centre, such as

[1] Further distinctions are sometimes drawn, such as *epistemic naturalism*, according to which only the sciences are capable of yielding knowledge, and *metaphilosophical naturalism*, which holds that philosophy is continuous with science (see, e.g., Beale, 2020; Glock, 2022).

[2] This ontology is most commonly referred to today as 'physicalism'. The term has largely replaced 'materialism', as the latter implies a world composed of matter in a sense that is no longer compatible with modern physics. Unfortunately, the term 'physicalism' is also problematic for present purposes, as there is little reason to believe that fundamental physics excludes the mind or its significance.

idealism.[3] If naturalists did not operate on these assumptions about the convergence and authority of science, they would have no obvious reason, on the basis of their core commitments, to reject idealism outright.

Contrary to what many naturalists, and even some of their critics, assume, it is far from clear what the sciences actually say about the material world. While there has been extensive debate in the philosophy of science about whether we should be realists concerning theoretical entities and natural laws, these discussions typically focus on the divide between realism and instrumentalism. That is, they ask whether scientific theories and laws are genuinely referential or merely tools for organising and predicting data.[4]

Yet the question of whether the mind plays a fundamental role in reality is a different matter altogether, and one that is typically sidelined in contemporary philosophy of science. Even if one is a scientific realist with respect to laws, it remains an open question what these laws are actually *about*. A Hegelian absolute idealist, for instance, would take them to describe spiritual or mental things. They would be the laws that the spirit follows in its development. More generally, even if arguments in the philosophy of science succeed in justifying belief in scientific entities under realist assumptions, they do not address whether such entities exist independently of the mind. For reasons that are likely as much cultural and psychological as philosophical, many contemporary naturalists seem to treat the traditional metaphysical debate between realism and idealism as settled, as though the sciences have definitively ruled out one side. But this is simply not the case. In fact, there is far less consensus among scientists about these matters than many philosophers presume. While some believe that "all fundamental things are now known and there is nothing left for us to do but fill in details," Nobel Prize-winning physicist Robert B. Laughlin counters that "[t]his [view] pushes my experimental colleagues beyond their already strained limits of patience, for it is both wrong and completely below the belt" (Laughlin, 2005, p. xiii). At this stage in scientific history, there is no compelling reason to assume that a more advanced and unified science will not ultimately favour a metaphysical picture in which the human mind, or some form of mind-like structure, plays a central role. Idealism, in some form, may yet return through the very channels that once appeared to dismiss it.

Many scientists regard the scientific image of the world as far less inimical to the mental than many philosophers do. Consider fundamental physics. Although

[3] The philosophical literature is notoriously vague about what qualifies someone as an 'idealist'. Kant, for instance, has traditionally been categorised as such, though many have reclassified him as a realist due to his acceptance of a noumenal world. In rare cases, even Berkeley is sometimes labelled a realist (see, e.g., Saporiti, 2004). Meanwhile, the neo-Kantian Hilary Putnam is often described as an 'anti-realist', a term no less ambiguous than 'idealist'.

[4] The instrumentalist view already finds support in Wittgenstein's *Tractatus* (1921), as well as in later work by the logical positivists (e.g., Schlick, 1948, the American pragmatists (e.g., Dewey, 1900), and also in the writings of Quine, 1963). While the instrumentalist allows for a subjective component in scientific theorising that might be influenced by social norms, values, or psychological tendencies, this view does not entail an ontological claim about the centrality of the mind.

some metaphysicians treat fundamental physics as offering strong reasons to be realists about self-individuating simples that compose macroscopic objects, quantum mechanics has traditionally paid considerable attention to the status of the observer (see, e.g., Bell, 1990; Wigner, 1967). While quantum mechanics does not *explicitly* incorporate the mind, it may nevertheless turn out to offer our best ontological option. Given that the current state of quantum mechanics arguably fails to provide a mind-independent ontology (Maudlin, 2019), and that many physicists regard it as a merely formal apparatus rather than a literal theory of the world—as captured by the popular slogan "Shut up and calculate!" (Mermin, 1989)—the idea that reality is shaped by the mind is not as implausible as it might first appear. Some have therefore argued that *relational quantum mechanics*, one of the leading interpretations of the theory, supports a broadly Kantian or idealist outlook (Bitbol, 2010). Moreover, Werner Heisenberg himself claimed that quantum mechanics imposes real constraints on the notion of a mind-independent reality. As he put it: "reality varies, depending upon whether we observe it or not" (Heisenberg, 1958, p. 52), a view he elaborated as follows:

> In classical physics science started from the belief – or should one say from the illusion? – that we could describe the world or at least parts of the world without any reference to ourselves ... [In contrast, quantum mechanics suggest that there exists] a subjective element in the description of atomic events, since the measuring device has been constructed by the observer, and we have to remember that what we observe is not nature in itself but nature exposed to our method of questioning ... in the language that we possess ... (Heisenberg, 1958, pp. 55–58)

While one might criticise Heisenberg for being indecisive about whether it is *reality itself* or rather its *description* that varies, this lack of distinction reflects a broader influence of late nineteenth- and early twentieth-century Kantian thought, which was widespread among many scientists of the time. In any case, if naturalists were to take science seriously, then physics would offer little justification for dismissing idealism out of hand.

Also within the cognitive sciences, some researchers have expressed sympathy for anti-realist ideas. For instance, the philosophers and scientists Francisco J. Varela, Eleanor Rosch, and Evan Thompson (VRT) (1991) raise the possibility that colour perception may offer evidence for a radically anti-realist position grounded in neuroscience. They argue that the limits of realist commitments to a mind-independent, objective world are best revealed through an analysis of how humans experience colour. The human retina contains three types of photoreceptor cells, known as cones, each sensitive to a different range of wavelengths in the visible light spectrum. Depending on the relative activity levels of these three cone types, a particular signal is transmitted to the next stage of the visual system. This interaction forms the biological basis for how Homo sapiens experience colour.

VRT use this insight to advance a central claim: since the same ratio of cone activity can be produced by different combinations of wavelengths striking the retina, distinct external states can give rise to identical perceptual experiences. Our colour experiences, therefore, do not stand in a one-to-one correspondence with properties of the external world. This implies that our understanding of the world is constrained and shaped by our biological constitution. If colour experiences do not directly reflect mind-independent features of the world, then colours, as we conceive of them, are not real properties of external objects, but mental constructs.[5] John Locke referred to such perceptual features as *secondary qualities*.[6]

If we read VRT in this way, their position is often said to be problematic: on the one hand, it has been criticised as incorrect; on the other, it is said to risk triviality. Realists object that VRT mistakenly infer a direct logical connection between the properties of the world and our human experiences of those properties. At the same time, they argue, it is trivial to observe that our description of the world is shaped by the particular way we are biologically constituted. Pigeons and goldfish, for instance, possess more than three types of cones in their retinas, and it is entirely uncontroversial to assume that they perceive aspects of the world inaccessible to us. Even if one grants that colours are secondary qualities, the realist maintains that the more significant question is whether *all* properties of the world are to be treated as secondary in order to substantiate the claim that the world is a mental construction. But such a sweeping conclusion, they argue, lies well beyond what VRT are capable of establishing (Shapiro, 2011, p. 55).

Demonstrating that *some* of our experiences are shaped by cognitive faculties is very different from showing that *all* of them are. Many critics demand from the idealist the stronger claim—that the entire structure of experience must be mind-dependent—in order for a view to qualify as idealist. But whether this demand is fair is questionable. It would imply that even if cognition were to generate an entire experiential universe distinct from what is really out there, the mere presence of a single property corresponding to a mind-independent world would disqualify the view from being idealist. This sets an unreasonably high bar and risks dismissing meaningful forms of idealism that acknowledge some structural overlap with external reality.

For the realist metaphysician, it is also unclear whether the approach taken by VRT still belongs to metaphysics, or whether it merely confuses metaphysical inquiry with questions in perception or epistemology. While we must surely have some form of access to the world for the realist position to make sense, this does not entail that the mind always delivers a perfectly accurate representation of how things are. A realist might simply invoke certain normality conditions to delineate the circumstances under which we can be said to access a mind-independent world, and would deny that all experience is a mental construct.

Nevertheless, however we interpret the examples of VRT, the broader point still remains: idealist tendencies can and do arise from within the sciences, and the results of science can often be interpreted in more than one metaphysical direction. Given the lack of consensus among scientists regarding the role of the mind in constituting reality, it remains entirely possible that large portions of science may one day embrace a metaphysics in which the mind is taken as ontologically prior

[5] This interpretation of VRT is discussed in Shapiro (2011, pp. 82–86).

[6] See Locke (1847) and Rickless (1997) for discussion.

to the material world. For these reasons, idealism and naturalism (at least in its metaphysical form) should not be regarded as mutually exclusive. To better grasp the stakes of this debate, I will now clarify how I understand the positions of realism and idealism as they will be discussed throughout the remainder of this book.

7.2 Classical Metaphysical Realism

Classical metaphysical realism is commonly defined by the idea that there exists a unique, mind-independent reality. Most versions of the view also assume that we are capable of describing this reality from an *Archimedean standpoint*, that is, from a perspective external to human thought and language.[7]

At its core, classical realism is committed to several ontological assumptions. First, it posits a world that exists independently of our conceptual schemes, linguistic practices, or mental states. The totality of things, or objects, that constitute this world exist irrespective of whether they are observed or represented by any mind. Facts about these entities are presumed to obtain regardless of our awareness of them. A further assumption often made, either explicitly or implicitly, is that the true ontology of the world encompasses many of the objects we encounter in everyday life as well as those posited by mature scientific theories (Horgan & Potrč, 2008, ch. 1). On this view, not only do trees and tables exist, but so do quarks and Higgs bosons.

In addition to its ontological commitments, realism typically involves an epistemic dimension. The world is said to be at least in part cognitively penetrable to us: we can form beliefs about it, and some of those beliefs can be true. Realist truth is usually conceived as correspondence—statements are true if they correctly represent or map onto the way the world actually is. On this view, truth requires reference and factual alignment between thought and reality.

Some critics argue that one could still be a realist while denying that we have any epistemic access to the world—for instance, by claiming that a mind-independent world exists even if we are cognitively confined to illusions or simulations. But such a position is arguably not metaphysically realist in the intended sense, as it fails to establish a meaningful link between ontology and epistemology. If all our beliefs are entirely disconnected from how reality actually is—as in a radical sceptical scenario like *The Matrix*—then we cannot claim to be engaging with reality at all. In that case, realism loses its philosophical purpose. For these reasons, classical metaphysical realism is typically committed not only to the existence of a mind-independent world, but also to our potential (if partial or fallible) access to that world.

---

[7] Many realists reject this stronger version of realism and acknowledge the epistemic limits of human cognition. Anti-realists often invoke the implausibility of such a standpoint to argue against realism altogether.

Based on these assumptions, the classical Quinean metaphysical realist sets herself the task of identifying those things that truly exist in the world, supported by the findings of the sciences. In previous chapters, I introduced the Quinean approach to metaphysics and explained how it is rooted in realist assumptions about scientific objects. I also noted that Quine himself might not—depending on interpretation—entirely endorse this approach. Ironically, Quine has not only been read as diverging from his own framework, but has even been described as a 'transcendental philosopher' (Veatch, 1978)—a label typically reserved for thinkers who rely on the methods associated with Kant's idealism. While characterising Quine as an idealist is certainly excessive for most interpreters, the extent to which he ought to be regarded as an anti-realist about material objects remains a matter of debate. Even though Quine, often seen as the father of modern metaphysics, describes himself as a proponent of 'robust realism' with an 'unswerving belief in external things—people, nerve endings, sticks, stones' (Quine, 1981, pp. 1–2), the degree to which he should be classified as a realist is still contested.[8]

Quine's realism is grounded in his naturalistic rejection of first philosophy, leaving science as the only justified standpoint for determining what exists and how we come to know it. Since the sciences posit entities at various scales, he does not adopt an eliminativist stance toward common-sense material objects.[9] Moreover, the fact that scientific practice must begin within the phenomenal world of immediate experience necessitates the postulation of manifest objects: 'Analyze theory-building how we will, we all must start in the middle. Our conceptual firsts are middle-sized, middle-distanced objects ...' (Quine, 1960, p. 4). Despite Quine's ontological realism regarding material objects, his epistemological commitments already introduce tensions with the broader metaphysical realist framework.

The traditional realist's theory of truth suggests that whenever a statement's predicate corresponds to the mind-independent object to which the statement refers, then the statement is true. For such a theory, the notion of reference that builds the connection between world and semantics is typically said to be required. If our judgements are to match the states of affairs obtaining in a world beyond our mind, then we depend on determinate referential relations between the words we use and the objects taking part in the states of affairs which ought to correspond to the sentences we make (Loux, 2006, p. 282). Thus, while some would say that a correspondence theory of truth might not be necessary for realism, if one conceives of the correspondence at issue in terms of the predicate corresponding to the object, then debates about reference gain importance. For this reason, the realist's reliance on a determinate notion of reference has been one of the more prominent worries about realism within the analytic tradition. For instance, if somebody were to assert the one-word sentence 'Rabbit,' it is not by mere ordinary usage of the term clear

[8] Davidson disagrees with Quine in light of such a non-realist reading; for discussion, see Glock (2003, ch. 5).

[9] It is, however, not always clear what Quine means by 'the sciences' when discussing his naturalistic commitments (Haack, 1993).

whether the sentence refers to brief temporal parts of rabbits or, alternatively, to undetached rabbit parts. Or yet again, the term might be equivalent to 'a portion of the single, discontinuous, yet unified object of all rabbits that have ever existed and ever will exist.'

According to Quine (1960), reference is only determinate *relative* to a background language or translation manual. This is, of course, closely in line with Quine's mentor, Carnap, and again seems to imply a form of anti-realism. For Quine, objects are in some sense human constructs that help us conceptualise empirical findings:

> Our talk of external things, our very notion of things, is just a conceptual apparatus that helps us to foresee and control the triggering of our sensory receptors in the light of previous triggering of our sensory receptors. The triggering, first and last, is all we have to go on. (Quine, 1981, p. 1)

According to Quine, what matters for a scientific theory is not the objects we postulate but rather how things relate to one another—how the theory is structured (Keskinen, 2012). All questions concerning the underlying nature of these things ought to be discarded, as they are products of our conceptual activity:

> The very terms 'thing' and 'exist' and 'real', after all, make no sense apart from human conceptualization. Asking after the thing in itself, apart from human conceptualization, is like asking how long the Nile really is, apart from our parochial miles or kilometers. (Quine, 1993, p. 113)

As we have already seen from the deflationary approach, making statements relative to a background language leads to the problem of determining and anchoring that background language itself, such that it becomes a mystery how different languages or frameworks relate to one another. And as we have seen, this again leads to the problem that we have no appropriate answer to the *clash*.

By stating that the subject has its own conceptual scheme which takes priority over other *possible* schemes, the indeterminacy of reference can, according to Putnam, be denied:

> Signs do not intrinsically correspond to objects, independently of how those signs are employed and by whom. But a sign that is actually employed in a particular way by a particular community of users can correspond to particular objects *within the conceptual scheme of those users*. 'Objects' do not exist independently of conceptual schemes. *We* cut up the world into objects when we introduce one or another scheme of description. Since the objects *and* the signs are alike *internal* to the scheme of description, it is possible to say what matches what. (Putnam, 1981, p. 52)

According to Putnam, reference is not directed toward the realist's mind-independent objects but rather toward the objects of our conceptual scheme, whose meaning is determined by their *use* within a linguistic community.

There is, however, a general problem with both Putnam's and Quine's arguably anti-realist positions. The concerns they raise are, from the realist's point of view, not metaphysical in nature but rather epistemological or linguistic. Faced with Quine and Putnam, the realist might simply nod and respond: 'Ok then, so we have a variety of ways to refer to the world. But what's the big deal? Ultimately, these are merely language issues. They have no bearing on my position as a metaphysical realist who believes that our task is to uncover the true structure of reality. There may be different linguistic ways of formulating such a theory and describing these structures, but in the end, all of them are concerned with describing what is out there, even if they do so in different ways. Just as different maps can represent the same terrain, different languages or theories may offer various perspectives on the same underlying reality.'

7.3 Modern Realism

While there are many variations of realism, the position ultimately maintains that we have, in one way or another, access to a mind-independent reality. In modern times, metaphysical realism has sometimes been interpreted somewhat differently from its classical formulation.

Based on deflationist considerations, existence questions have moved more into the background.[10] As we saw in the introduction to this book, metaphysical realists have always set themselves the task of identifying the general structure of reality. However, in the modern conception, existence is not always regarded as an adequate structural notion for describing reality, and new forms of structure are sought instead. The metaphysical realist Ted Sider (2011, ch. 1) illustrates these ambitions through a simple thought experiment.

Imagine a universe entirely filled with two liquids, one grey and one black. The liquids are neatly separated into two distinct halves, as illustrated in Fig. 7.1a. Carving reality at its joints is not a matter of arbitrary description. In the case of

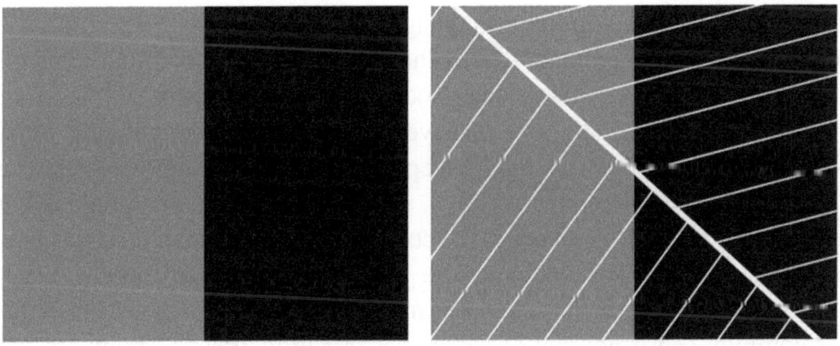

(a) (b)

Fig. 7.1 (**a**) Grey-black world, (**b**) Bizarre carving of the grey-black world. Adapted from Sider (2011, pp. 1–2)

[10] Consider, for instance, the approach of identifying grounding relationships between entities, which has also been argued for within the Quinean tradition (Schaffer, 2009).

the grey-black world, there is a single best way to describe the situation: by dividing the world with a plane that separates the black substance from the grey one. Now imagine a group of metaphysicians who divide the world differently, as marked by the thick white line in Fig. 7.1b. The region to the left of the white line is what they call 'glack', and the region to the right is 'bley'. Are the predicates introduced by these metaphysicians really those that carve reality at its joints? Most of us would think not.

Their mistake is not one of misidentifying the location of the grey and black regions, since they are not making any claims about grey or black. Nor are they wrong in saying that the region to the left of the dashed line is 'glack' and the one to the right is 'bley'; these are their own engineered concepts, and applying them in this way follows from their definitions. The problem lies in the very concepts they are using. They have introduced terms that fail to carve reality at its joints, despite their belief that 'bley' is bley and 'glack' is glack are true. Their conceptual scheme simply does not match the structure of the world.

The second modification of metaphysical realism concerns its epistemic demands. The critiques of realism advanced by Quine and Putnam require an adaptation of the classical realist's understanding of reference and its associated commitments. While these objections are historically significant, current developments in both science and philosophy offer even stronger reasons to revise the classical realist picture.

If one holds that the organisation of the mind is best understood in terms of a language-like system—*the language of thought*indexlanguage of thought—that contains symbolic representations referring to objects in the world, then it seems natural to treat the brain as a kind of language system. The symbols within the brain would be *about* objects in the world, in much the same way that noun terms in natural language refer to external things. However, this depiction of the relationship between mind and world begins to shift if we assume that the brain functions in a way that is fundamentally unlike a language system.

So why should the brain not be a language system containing symbols that are about the world? At first glance, such a claim may seem outrageous, since our everyday experience suggests that we are constantly manipulating words and symbols and relating them to our perceptions. Yet, as radical as it may appear, the eliminativist Alex Rosenberg (2011, ch. 8) argues that we should accept the inscrutability of reference and reject the common-sense idea that thoughts are *about* things in the world.[11] He dismisses the idea that a clump of matter—such as a brain—could be *about* another clump of matter, and contends that there is no room for our common-sense notions of reference, intentionality, or aboutness within a scientific worldview.[12] While the brain may receive, store, and process information,

[11] Note that Rosenberg's views are not grounded in the arguments of Quine or Putnam, but emerge directly from neuroscientific findings.

[12] Others, such as Kenny (1984, ch. 9), Bennett and Hacker (2003), and Maslin (2001), share this general sentiment but draw different conclusions. According to them, cognitive processes should

it cannot, Rosenberg claims, refer to objects in the external world. To make this argument plausible, he discusses an experiment by Nobel laureate Eric Kandel investigating the neural basis of classical conditioning.

In classical conditioning experiments such as Pavlov's dog, an animal is repeatedly presented with an unconditioned stimulus (e.g., meat), which triggers a natural behavioural reaction, together with a second, neutral stimulus (e.g., the ringing of a bell). Over the course of the learning process, the second stimulus eventually elicits the same behavioural response (e.g., salivation) as the first. This experiment was foundational to behaviourism, as it appeared to show that internal states in the animal's mind had a clear manifestation in observable behaviour. It might seem, therefore, that the animal possesses mental states that are in some sense *about* the world.

In the 1960s, however, Kandel initiated a research programme aimed at uncovering the neuroanatomical basis of learning. His goal was to understand what the nervous system was doing, at the neurological level, during conditioning. To this end, he began with experiments on a comparatively simple organism: the sea slug, which possesses only around 20,000 neurones—a small number compared to the estimated 70 to 100 billion neurones in the human brain. In these experiments, the slug's nervous system was exposed to two types of electrical stimulus: one harmless, the other painful (inferred from the gill-withdrawal reflex). The two stimuli were repeatedly applied in combination, until eventually the harmless stimulus alone triggered the same reflexive withdrawal as the painful one.

What Kandel and his colleagues discovered was that learning involves changes in the concentration of certain molecules, so-called *neurotransmitters*, at the synapses (Carew et al., 1971). This biochemical discovery supported Hebb's (1949) theory of synaptic plasticity, according to which changes in the strength and number of synaptic connections between neurones underlie memory. These findings were later shown to apply to mammalian brains as well, and there is no reason to believe that human brains function differently in this regard.

According to Rosenberg, this shows that, contrary to intuition,

> the sea slug does not actually learn and store any information that could be expressed in thoughts *about* stimuli ... What the sea slug has learned is not some new fact *about* the world. It has acquired a new habit to do something under certain conditions. It did this because its neurones were rewired into a new input/output circuit, treating the new habit ... It's learned to do something; it hasn't acquired a thought *about* something. (Rosenberg, 2011, p. 146)

What appears to us as a fundamental relational predicate Rxy, connecting two objects, is, from Rosenberg's scientific perspective, an illusion. The brain does not encode a *map* of reality that is *about* the world. Rather, Rosenberg concludes that the mind–world relationship must consist in a sameness of form and structure, requiring

not be attributed to the brain in isolation, but to the person as a whole. Still others, like Lycan (1991) and to some extent Dennett (1975), who defend versions of homuncular functionalism, conclude that the brain is not merely a clump of matter.

no further *interpretation*; the mind is, in his view, *physically isomorphic* to the world.[13] If the world stands in an isomorphic relation to the brain, then structural changes in the world result in corresponding structural changes in the brain.

This leads to a revised formulation of the correspondence theory, according to which neural representations correspond, more or less closely, to distributions of matter and energy in the environment (Rosenberg, 2017). However, one must note that the term 'representation' here no longer implies that there exists an entity 'x' in the brain that symbolically represents an object 'x' in the external world. Rather, it refers to the connectivity matrix of weighted synaptic links between neurones being structurally isomorphic to patterns in the world. It does *not* imply the existence of a mental object M with the symbolic content 'x' that refers to an external object x beyond the physical boundaries of the organism. In this sense, whatever representation the sea slug has of the world is of a fundamentally different kind from the notion of reference traditionally attributed to human cognition by realist philosophers.

Taking Rosenberg's views as prototypical of a science-oriented realist position, we can summarise the modern realist conviction as follows: there exists a mind-independent, discourse-independent totality of things, along with structural relationships between those things. Since the mind stands in an isomorphic relationship to this totality, we can aim to understand these structures. The notion of existence, however, does not seem to offer the right conceptual tool for carving these relationships.

7.4 Idealism

Idealism is often referred to as an extreme form of anti-realism, although the distinction between the two is not always clear. Since idealism is today a rather fringe position, there is no generally accepted way of articulating it. In contrast to my discussion of realism, where I attempted to outline a mainstream attitude among philosophers, I will here largely present my own way of thinking about the idealist viewpoint.

Many philosophers associate idealism primarily with the mind–body problem, that is, the question of how the mind relates to the body or to the material world more generally. In these discussions, idealism is often taken to be the view that only the mind truly exists, and that matter is either a representation or a mental construct. This, however, is not the understanding of idealism I adopt here. Instead, I follow Thomas Hofweber in broadening the notion to concern our general conception of reality:

[13] Technically speaking, an isomorphism is a structure-preserving mapping that can be reversed by an inverse mapping.

> Idealism does not have to be understood as being primarily concerned with the relationship between minds and matter. It should instead be seen more broadly, as being concerned with the place of minds in reality. (Hofweber, 2019a, p. 701)

If, as I have argued in previous chapters, making sense of reality as a whole is a project of unifying and resolving the clash, then idealism is the claim that only by including the mind can this project succeed. It is the view that the mind must be taken as central to any adequate account of reality.

What does idealism mean more precisely? Most philosophers who discuss reality define it either as the totality of all things or as the totality of all facts. These two notions are clearly related: for any x to exist, there also exists at least the fact that x exists, and vice versa. Nonetheless, as Hofweber (2019a) argues, it is useful to distinguish between *ontological idealism* and *alethic idealism*. According to the former, minds are central to reality understood as the totality of things, whereas the latter holds that minds are central to reality understood as the totality of facts.

Let us focus on reality understood as the totality of facts. An example of a fact is that snow is white. In this case, 'snow' is the object and 'white' is its property. Another fact is that most swans are white, which shares the same structure as the previous fact and also points to a similarity between the objects involved. Whatever fact we formulate, it appears to follow a specific structure, namely, an object–property structure. This structure also seems intimately connected to our thoughts and to language. The fact that snow is white is conceptually captured in thought and language by the sentence 'snow is white'. Both thought and language exhibit a subject–predicate form, Px, where x is a subject such as 'snow' and P is a predicate such as 'white'. It seems that the structure of facts corresponds closely with the structure of our thoughts. There appears to be a unified structure, call it the *Px-structure*, that links language, thought, and reality. But the question is: why?

As Hofweber writes:

> Naturally, there is a connection between the thought and the fact it is about. The form of the thought seems to match up perfectly with the structure of the fact. A subject-predicate thought represents a fact with an object-property structure. ... But why is there this match? ... The realist will hold that our thoughts have their form because the facts have the corresponding structure. And an idealist can see it the other way round: the facts have the structure because our thoughts have the form. (Hofweber, 2019a, pp. 704–705)

Do our thoughts and language mirror the facts because that is how reality is, and we have access to it? Or is it the other way around—do our thoughts have this very structure, which we then project onto the world? Figure 7.2 illustrates this fundamental question about how language, thought, and reality align.

The realist believes that there is an isomorphic relationship between reality and thought because humans have access to reality through their cognitive capacities. The idealist sees things differently. They maintain that the facts in the world have this particular structure only because they conform to the structure of our thoughts. If our thoughts were structured differently, then the facts we discover

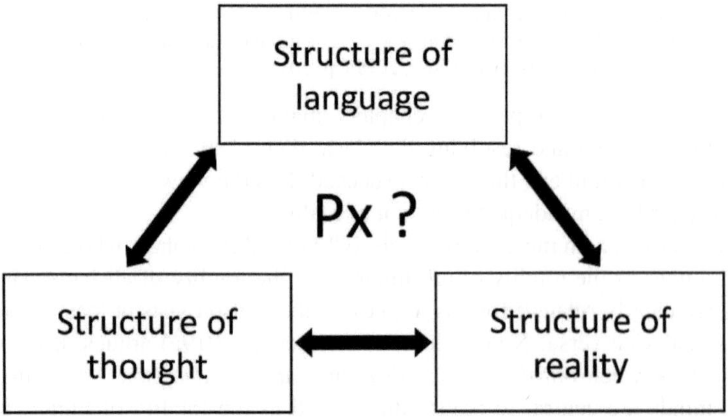

Fig. 7.2 Structures of thought, language and reality all appear to have the same structure Px whereby P is either property or predicate and x is either object or subject, depending on whether we are talking about facts or sentences about facts

about the world would also exhibit a different structure.[14] The idealist thus sees the apparent structural match between thought and world not as evidence of a genuine isomorphism, but as the result of a mental projection. In their view, the structures we attribute to the world are imposed by the mind, and the impression of an isomorphic relationship with a mind-independent reality is ultimately illusory.

There are, of course, more structural aspects to reality than the Px-structure. While the Px-structure is particularly significant, being the most basic and shared across all facts, other structural aspects exist as well. One example is the relationship between an object and its parts. Even if the description of wholes and parts depends on the Px-structure, the whole–part relationship may have different origins than the Px-structure itself.

If we define idealism as an absurdly extreme view, such as the position that *all* aspects of reality are mental, then the debate between idealism and realism will inevitably tilt in favour of realism, since one would have to embrace solipsism or a similarly radical stance to defend idealism. If, however, we frame the debate not as an all-or-nothing dichotomy but as a continuum—with the most radical versions of each position at the extremes—then a more productive and nuanced discussion becomes possible. This continuum concerns the degree of mental projection involved in our experience of the various images of the world.[15] Unfortunately,

[14] Hofweber argues for idealism by showing that realism depends on the existence of ineffable facts. According to his account of how language relates to the world, such facts cannot exist—which is why he favours idealism. I will pursue a very different route that does not focus on language.

[15] Accordingly, one might adopt an idealist stance regarding the manifest image, while maintaining a realist position with respect to fundamental physics.

it is precisely this nuance that is often lost in the literature, where the debate is frequently presented in strictly binary terms. Kant's philosophy exemplifies these difficulties particularly well. In some respects, he adopts a realist position, but when it comes to certain essential structures, he is also an idealist. This refusal to align himself fully with either extreme is what makes his position so intriguing to many philosophers:

> Kant was not the sort of person who had the intellectual courage to face up to a dilemma and reject one alternative in favour of the other. Instead, when he felt himself pulled in opposite directions by conflicting imperatives, his preference was to try to work out some way of satisfying them both. This intellectual cowardice ... is the characteristic that led Kant to his most brilliant discoveries. (Falkenstein, 1995, p. 19)

What, then, should we do with our philosophical taxonomy if, for instance, the Px-structure were a product of the mind, while other structures originated in a mind-independent world—or vice versa? Given the historical development of debates in the field, and the widespread commitment to a comparatively strong realism in contemporary metaphysical literature, I will use the term *idealistic philosophy* to refer to any view that takes the Px-structure, along with most mereological relations and other central metaphysical structures, to be the result of how the mind operates. These further structures typically include the non-collocation of objects, persistence through time and space, vagueness, and principles of compositionality. It is no coincidence that these are the very structures most involved in classical metaphysical puzzles: they often do not reflect how the world is, but rather how the brain presents the world to us—at times leading us into inconsistencies when we assume these structures to exist independently of us. For the sake of brevity, I refer to these as the *essential structures* of material objects.[16]

An idealist need not hold that *all* structures are mental in origin, but only that those most crucial to our philosophical understanding, especially those invoked in explaining the puzzles of ordinary objects, are mind-dependent. On this view, idealistic philosophy reveals itself by accounting for such puzzles in terms of cognitive structures, rather than by appealing to the nature of a mind-independent world. The presented definition is, of course, somewhat arbitrary, but it reflects the necessity of drawing a conceptual boundary along what is, in truth, a continuous spectrum between idealism and realism. Some structural features may be mentally constructed, others not, and the task is not to take sides absolutely, but to locate ourselves thoughtfully within this continuum.

[16] Notice that all ordinary objects are necessarily structured in accordance with these essential structures, whereas the objects of physical theorising typically lack features such as compositionality and vagueness. How this structural clash arises will become more apparent throughout the book.

7.5 Eliminativism and Idealism

Ironically, when it comes to the manifest image, idealism is surprisingly close to radical eliminativism, insofar as both agree that the manifest image is, to a large extent, a mental construct. However, the eliminativist is a realist with respect to the objects of fundamental physics and sees no metaphysical relevance in the manifest image. As we have seen, eliminativism suffers from the problem of failing to provide a useful framework for most of reality, including ordinary objects such as tables and chairs, as well as non-fundamental scientific entities. This is part of the motivation behind Ladyman and Ross's attempt to introduce an alternative unifying proposal, which, as Harman explains, fails when interpreted in a realist framework.

The difficulty with Ladyman and Ross's account lies in its attempt to preserve the Px-structure for their so-called real patterns as individuals of some sort while also upholding realism. Their position assumes that real patterns appear as a plurality at each level of reality, distinct from other levels, but it does not address questions concerning the essential structures of objects. The fact that the fundamental world is structured very differently from the manifest image of the world may, in fact, motivate a form of idealism about the world we encounter in everyday life.

But is this shared idealist view of the manifest image not obviously true, given how different fundamental reality is from our manifest world? Not quite. A realist might respond that, within the manifest image, the Px-structure is indeed a reflection of how the world really is. Such a realist might defend their position by saying something like: 'It is true that, at the fundamental level, the world is structured very differently from the world of appearances. But when we are dealing with the manifest image, it is unfair to compare it directly with the structures from other parts of reality. Perhaps it is still the task of physics to discover how, at macroscopic scales, physical principles reach a limit that is best characterised in terms of essential structures.'

It is at least reasonable to assume that metaphysicians who engage in debates about objects within the manifest image—debates about statues and lumps of clay, the vagueness of objects, compositionality, and so on—must be committed to some version of this realist response. Otherwise, their engagement in metaphysics would be difficult to justify.

To motivate the form of idealism proposed in this book, we must therefore first address this realist concern by identifying reasons why there is no isomorphism between our mental representations and the world, even when we restrict ourselves to the manifest image. Only then can the naturalist's unification programme be transformed into an idealistic metaphysical project—one that avoids the problems inherent in realism.

Summary
Ontological naturalism and idealism are not mutually exclusive. In fact, there are several scientifically motivated reasons to support a form of idealism. The version of idealism explored in this book centres on the essential qualities of material objects,

which underlie a range of metaphysical puzzles. According to the definition adopted here, one qualifies as an idealist if one holds that most of these qualities do not correspond to (in the sense of not being isomorphic to) any actual features of the world.

Chapter 8
Imposing Structure onto the World

We have seen that many metaphysical arguments draw support from our ordinary intuitions, which are shaped by experience and by the way we navigate the world. This, however, gives rise to a serious concern. At the fundamental level, physics operates with concepts that are far removed from those rooted in everyday understanding. As a result, metaphysical attempts to relate the fundamental structure of reality to the macroscopic world often struggle to gain traction. Consider expressions such as 'tables are particles arranged table-wise'. These are attempts to bridge the gap between levels of reality, but they fail to do so because the conceptual frameworks used to describe one level are often incompatible with those used at another. Still, these levels must be connected, since they are part of a single, unified reality.

The radical eliminativist responds by suggesting that we simply discard the objects of the manifest world from our ontology. By contrast, the liberal naturalist is right to defend their place. This raises a central question: even if the objects of our experienced world differ fundamentally from the entities posited by physics, might it still be reasonable to believe that our way of thinking about the manifest world reflects how things truly are at that particular scale? Most people would be inclined to say yes: the way we structure reality at the macroscopic level is broadly correct, and the real difficulty lies in connecting the various levels. On this interpretation, the concepts we use in everyday experience accurately capture that level of reality. More generally, each level might be said to admit of a true description in terms appropriate to it.

In the final part of this book, we will see that this outlook is supported only by idealism, not by realism. For now, however, our focus will remain on the manifest image. This chapter will argue that if we can explain how the brain gives rise to the perceptual structures that define our lived world, then the realist is poorly positioned to claim that these structures exist independently of the mind. Whether neuroscience can ultimately provide such an explanation will be the concern of later chapters. What will also emerge from the development of this argument is a

A. C. Schwaninger, *Cognitive Metaphysics*, Synthese Library 529,
https://doi.org/10.1007/978-3-032-05420-3_8

novel methodology. It builds on a naturalised reinterpretation of Kant's idealism. We will see that, when read in a certain way, Kant offers a compelling starting point for a naturalistic approach to metaphysics grounded in the tools of computational neuroscience.

8.1 Evolution and Veridical Representation

Many metaphysicians believe that, ultimately, the questions they raise cannot be settled on empirical grounds, because metaphysical theories are underdetermined by empirical data (Ribeiro, 2015).[1] However, there is a major problem with arguments that portray metaphysics as being completely independent of human experience: it is only due to our interaction with the world and our experiences that we even have metaphysical concepts at our disposal. Consider, for instance, the concept of compositionality. The concept of a composing object would simply not be formed in our mental concept formation processes if no interaction with what we experience as composing objects had taken place (Hofweber, 2019a). From a realist perspective, making a claim about the mind-independent structure of macroscopic, material objects amounts to claiming that the world around us is pretty much structured the way we see it to be.

If we accept that the concepts that characterise ordinary objects are formed on an empirical basis, an urgent question arises: can we trust that this empirical basis is accurate, in the sense that the concepts metaphysicians employ in stating that basis are actually concepts that allow us to investigate the nature of reality (realism), or are these concepts merely the products of our innate neurological concept formation processes (idealism)? It seems that, when doing metaphysics about ordinary objects, we rely on the belief that the concepts formed by empirical means ought to be relevant in the most general explanations of the world, due to their universal applicability within the world of our daily experiences.

The metaphysical realist must rely on the assumption that our thoughts and the world stand in a 'special relationship' (Loux, 2006, p. 288) to one another, allowing us to form concepts that can be accurately applied to the natural world. It is typically assumed that the concepts we employ in metaphysics are shaped through interaction with the world for evolutionary reasons (see, e.g., Quine, 1969). While we may lack the biological constitution required to form the appropriate concepts to fully grasp objects existing at the quantum scale, the selection pressures faced by our ancestors have enabled humans to acquire the cognitive capacities necessary to form concepts that allow us to track the world with reasonable accuracy within our ecological niche and at our experienced level of reality.

---

[1] Some arguments are claimed to be completely independent of experience. For instance, one might argue on a priori grounds against restricted compositionality by pointing to the logical fact that it entails metaphysical vagueness (see, e.g., Lewis, 1986a, pp. 212–213).

This is confirmed by many, if not most, scientists, who tell us that we can trust our visual experiences to be veridical under certain normality conditions. From this perspective, it seems reasonable to take our senses as a good starting point in our inquiry into the true structure of reality. When we turn to classical accounts of vision, for instance, we find a clear optimism about humans having access to a true description of objective reality: 'usually our perceptual processing does run correctly (it delivers a true description of what is there)' (Marr, 2010, p. 30). This optimism about our senses is motivated by the idea that the fitness of a living system improves the more veridical its perception of the world is:

> Our sense organs have *evolved* to give us a marvelously detailed and accurate view of the outside world... our sensory systems are organized to give us a detailed and accurate view of reality, exactly as we would expect if truth about the outside world helps us to navigate it more effectively (Trivers, 2011, p. 2, emphasis added).

The realist must assume that, over the course of evolution, our minds have developed to account for the facts in the world by adopting a structurally corresponding representation. Biologically speaking, the realist holds that it would have been a waste of resources to maintain a representation of the world that diverges from how the world actually is. Evolution, on this view, should have led to structures in thought that correspond to the essential structures of the world. If the world were not structured in the way we perceive it to be, we would simply not survive—right? Not necessarily. Even though realist accounts of vision align with our intuitions,[2] there is nothing in the theory of evolution that requires us to accept the conclusion drawn in the previous quote, nor is it in any way clear that natural selection favours a disposition to perceive accurately.

> We are organisms, not angels, and our minds are organs, not pipelines to the truth. Our minds evolved by natural selection to solve problems that were life-and-death matters to our ancestors, not to commune with correctness or to answer any question we are capable of asking. (Pinker, 1997, p. 561)

Consequently, certain developments have pointed in a different direction from the standard, intuitive view of the world.

Donald Hoffman (2019) presents one of the most uncompromising challenges to the realist view. According to him, '[s]eeing truth hides fitness, and seeing fitness hides truth' (p. 148), which he illustrates with the following example (ch. 4). Imagine a creature called a *kritter* that lives off a resource called *stuf*. Just as humans cannot survive in environments with either too little or too much oxygen, there is an optimal amount of *stuf* with respect to a kritter's fitness level (see Fig. 8.1a). Also assume that a kritter has only two perceptions: grey and black.

Suppose a *Truth* kritter tracks reality as it truly is. It will perceive grey in regions of low *stuf* concentration and black otherwise, as illustrated on the left-hand side of Fig. 8.1b. On the right-hand side, we see what a *Fitness* kritter perceives—one that

---

[2] We should not forget that even a person living in the *Matrix* would find intuitions to be a good guide to her convictions, making intuitions an unreliable source of justification.

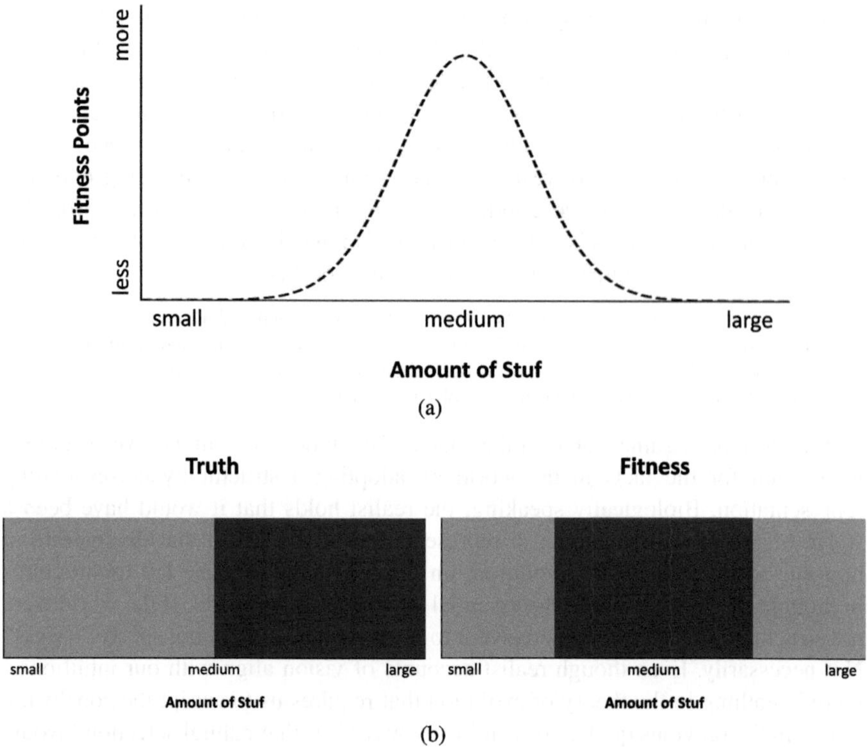

Fig. 8.1 (**a**) Fitness function of a kritter; (**b**) shades of grey as perceived by the two types of kritters. Adapted from Hoffman (2019, ch. 4)

has evolved to track fitness rather than truth. The *Fitness* kritter structures its world to reflect whether the right amount of *stuf* is available, not whether there is little or much of it. Which of the two kritters has the higher chance of survival? Clearly, the *Fitness* kritter; the *Truth* kritter is unable to determine whether its current behaviour is adaptive or not: 'fitness payoffs are distinct from objective reality and can, for a given element of reality, vary wildly from creature to creature and time to time. To track fitness is simply not, in general, to track truth' (p. 206).[3]

In line with this, Hoffman writes about material objects that

> Our perceptions of the moon and other objects were not shaped to reveal objective reality, but to disclose the one thing that matters in evolution – fitness payoffs. Physical objects are satisficing displays of crucial information about payoffs that govern our survival and reproduction. They are data structures that we create and destroy. (p. 176)

[3] For Hoffman, a veridical perception preserves the true structure of the world. It assumes an isomorphism—a structure-preserving map—between the world as it is, independent of an observer, and how we believe it to be.

An obvious objection to this example is to deny that the Fitness kritter is misrepresenting reality. One might argue that it is simply carving up reality differently—by perceiving all medium amounts of *stuf* as black, and all small and large amounts as grey. From this perspective, the choice of representation appears arbitrary, with both schemes being equally valid. However, let us recall the Grey-black world discussed in Chap. 7, which illustrates that some ways of carving up reality can in fact be worse than others.

In the case of the Truth kritter's carving, the transition from grey to black occurs at an arbitrary point, as is typically necessary for vague categories. However, in contrast to the Fitness kritter, the boundary still preserves a *continuity* in the change of *stuf* within a single perceptual colour. For the Fitness kritter, by contrast, the grey area contains a discontinuity that can be traced only to the properties of the kritter itself, and not to any property of the world.

Hoffman is right to say that fitness and truth can be in conflict, but as critics point out, one must resist presenting this as an either-or debate (Shermer, 2015).[4] Even if it is the case that all perception is entangled with subjective input, the information we obtain also contains data about an objective world—to a greater or lesser extent—even if the two components of input might never be fully separable. This implies that certain features of the experienced world might be more objectively truthful than others, and that a sharp distinction between the mind-independent world and the world of experience is misguided.

8.2 Kant's Metaphysical Methodology

At the very least, the foregoing considerations ought to prompt a measure of scepticism towards metaphysical realism. It is by no means evident that access to truth confers any biological advantage, and no known evolutionary mechanism appears poised to steer our cognitive apparatus reliably towards the truth. Doubts of this kind are not novel. Philosophers across the centuries have registered unease with the assumption that the conceptual toolkit of the mind grants us unmediated access to reality as it is in itself.

Kant's *Critique* is perhaps the most famous response to this problem. He was reacting to the tradition of *transcendent* metaphysics, which aimed to uncover the ultimate nature of reality by relying purely on reasoning that does not depend on experience—what philosophers call a priori reasoning.[5] Thinkers in this tradition believed that by using the mind alone, without reference to observation or experi-

[4] Note that others have criticised Hoffman for a misapprehension of the notion of veridicality (Cohen, 2015), as well as for his methodology in attempting to prove the radical claim that none of the structures we perceive in reality are real (Martínez, 2019; Mausfeld, 2015).

[5] This includes ancient views such as Platonism, which posits abstract entities beyond space and time, Aristotelian metaphysics concerned with 'being qua being', and many medieval theories about God.

ment, one could arrive at deep truths about the world. Kant, however, challenged this idea. His reasons are echoed by many contemporary metaphysicians: first, because there is no clear method for gaining reliable knowledge about the world in this way outside mathematics and logic; and second, because the tradition of transcendent metaphysics has failed to produce a stable, widely accepted body of results.

Kant's proposal is to redirect metaphysical inquiry by confronting a moderate but powerful sceptical challenge: that we lack access to the world *in itself*, and that metaphysics must therefore turn its attention to the conditions that make knowledge and experience possible (Glock, 1997, p. 287). In doing so, Kant does not reduce metaphysics to psychology or epistemology. Rather, he reimagines it as a 'metaphysics of metaphysics', that is, an investigation into the form such an inquiry must take if it is to yield genuine insight.[6] The task of metaphysics, properly understood, is not to disclose what *being* is in itself, but to elucidate the necessary preconditions for any possible knowledge of it. Kant names this project *transcendental* or *critical* metaphysics.

Kant observes that the objects we encounter in the world exhibit certain necessary and universally shared properties. They persist in space and time, stand in causal relations to one another, and appear as wholes composed of proper parts, while themselves being parts of larger wholes. But what, according to Kant, is the correct metaphysical account of these universal features? He argues that a proper investigation into the possibility of metaphysics leads us to uncover the non-empirical, yet metaphysical, preconditions of experience:[7]

> the conditions for the *possibility of experience* as such are simultaneously conditions for the *possibility of objects of experience* and hence have objective validity in a synthetic a priori judgment. (A 158/B 197)

These preconditions of experience determine the necessary properties of the objects of experience. These properties are *necessary* in the sense that experiences or mental imagery of objects are not possible for the cogniser unless the objects possess these features.[8] Since these properties are present throughout all our experiences and acts of imagination, we can analyse the categories of cognition that we impose upon the world and that make such modes of intuition possible. For Kant, '[w]e can have a priori knowledge by means of the categories, only if the categories are due to the nature of the mind and are imposed by the mind on the objects which it knows' (Paton, 1936, p. 258). By their a priori nature, Kant's categories apply universally to all objects of possible cognition (Kant A 79/B 105) and are not conditional on *Homo sapiens* (Strawson, 1987). Owing to their universal status, their significance extends not merely to perception and cognition, but also to ontology. As a result, the Aristotelian

[6] See Kant's Letter to Marcus Herz, 11 May 1781/X.

[7] I refer to Kant's (1781/1787) first and second editions of the *Critique* as A and B.

[8] These features are also necessary in the sense of being non-contingent on the biological faculties of the cogniser. I will, however, postpone this aspect to a later point in the chapter.

> investigation of being qua being is understood as an investigation into the essential features of objects, and hence into *what it is to be an object.* These essential features of objects are determined by the essential ways in which we experience or represent objects. (Glock, 1997, p. 290),

Despite the difficulty of having no, or only limited, access to the world as it is in itself, Kant offers a path for metaphysical inquiry that focuses on the nature of the objects of our experience by turning to cognition and investigating the limits of possible knowledge. The resulting propositions of his *critical metaphysics*, which he refers to as a 'pure natural science,'[9] are *synthetic* a priori: they are justified independently of empirical observation and are not mere logical derivations from concepts.[10] These metaphysical propositions express fundamental theoretical commitments that structure our sensory input and are therefore constitutive of the objects we experience.

The metaphysical propositions Kant seeks represent fundamental theoretical commitments that give structure to our sensations and are thus constitutive of experience in his sense:

> Experience is empirical knowledge, that is, a knowledge which determines an object through perceptions. It is a synthesis of perceptions, not contained in perception but itself containing in one consciousness the synthetic unity of the manifold of perceptions. (A177, B218)

The 'manifold of perceptions' corresponds to what Quine describes as the sum of our 'surface irritations, which exhaust our clues to an external world' (Quine, 1960, p. 22). According to Kant, the synthetic a priori propositions of his metaphysics determine the very conditions under which empirical inquiry is possible, and they ground the criteria by which we can be right or wrong in our analyses (Machina, 1972).

8.3 Naturalising Kant's Methodology

All empirical research takes natural objects (or events) as its subject matter, and scientists aim to discover laws that characterise the behaviour of these objects (or events). According to Kant, the very possibility of such investigation depends on the mind and the categories it imposes on that subject matter. To avoid circularity, Kant denies metaphysical significance to any attempt to investigate our cognitive structures through empirical psychology. The universal categories that shape all knowledge and experience must, for him, stand outside the empirical realm and transcend it.

---

[9] Kant uses the phrase 'pure natural science' (*reine Naturwissenschaft*, in German) (see, e.g., B 20).

[10] Synthetic a priori propositions fall into two groups: the propositions of arithmetic and geometry form the first group, and the propositions of critical metaphysics form the second (Machina, 1972). In the present discussion, I refer only to the latter, which concern the nature of ordinary objects.

Instead, Kant maintains that *transcendental arguments* allow us to identify the metaphysical categories that give structure to the reality we cognise, and, by extension, to the reality cognised by any rational being. These arguments proceed from an observed or agreed-upon feature of experience Y and aim to show that some condition X must hold in order for Y to be possible. Their structure is often summarised as: X is a necessary condition for Y; Y obtains; therefore, X must also obtain. While this may resemble a mere instance of modus ponens, transcendental arguments are intended to reveal *the conditions of possibility* for experience or knowledge, not to infer empirical facts from other empirical facts. Typically, Kant draws on introspective judgements and argues that certain structural features of cognition are required to make such judgements possible, from which he infers the universal status of those features.[11]

We might be hesitant to believe that Kant's metaphysics, which turns to cognition, could yield any deep insights, especially given how mysterious cognition already is and his reliance on a priori transcendental reasoning. Yet even from a naturalist perspective, there is good reason to take this approach seriously. A turn to cognition may offer a viable path to understanding the underlying structures of perception. Surely no naturalist would object to investigating the conditions for the possibility of knowledge, nor to the idea that these conditions are rooted in ourselves. Moreover, approaching questions about the nature of objects by treating singular terms as mental representations is not at odds with naturalist commitments. On the contrary, there are compelling reasons to prioritise mental over linguistic representations: the former are explanatorily prior and account for object-directed thought in infants and non-human animals (Strang, 2017).

While ontology is, to some extent, a linguistic matter shaped by our background assumptions, metaphysics as a whole reaches beyond the confines of language. What many naturalistically inclined metaphysicians do question, however, is whether the mind can be examined through a priori reasoning:

> So the basic invention of Kant, to distinguish between two levels of discourse, the empirical and the transcendental, conflicts with naturalism. It is precisely for this reason I prefer naturalism, since I have doubts about any a priori inquiry into our minds. (Johansson, 2011, p. 118)

To many, it remains doubtful whether an introspective method can deliver what Kant's critical metaphysics sets out to achieve.

Given that transcendental arguments are not meant to depend on empirical observations, their very possibility—and, with it, the consistency of Kant's project—has been called into question:

> It is not clear why Kant's own brand of what might be called 'transcendental psychology' – which is clearly distinguished from the empirical – does not fall under the same interdict that he places on rational psychology. (Strawson, 1987, p. 408)

[11] Transcendental arguments are typically used as a response to sceptical concerns.

Kant criticises the rationalist tradition for seeking a priori truths about the noumenal world, and yet he himself offers a methodology that relies on a priori reasoning. Is this not a straightforward contradiction?

Some exegetical work has shown that Kant successfully defends himself against this objection, thereby preserving the internal coherence of transcendental idealism. Even though we formulate transcendental arguments on the basis of certain experiences, 'we do enter the field of noumena, we do not enter it "in this way," that is, by a strict a priori justificatory process. Consequently, Kant can consistently deny our right of advancing yet further in the noumenal domain' (Pereboom, 1991, p. 370). In other words, although transcendental metaphysics tells us something that lies beyond the empirical, namely, it identifies transcendental categories, our access to this knowledge still depends on experience.

The notion of 'a priori' should therefore not be understood in a strict sense, as if all metaphysical reasoning required a completely non-empirical foundation. Kant's method does not forbid empirical reasoning altogether; rather, it urges us to identify universally shared features of experience and to infer from them the a priori principles that make such experience possible. We can accept that there is an a priori element in Kant's reasoning without concluding that it refers to objects beyond all possible experience.

To identify the necessary conditions for the experience of objects, it is unavoidable to make some empirical observations that help reveal the categories required for having knowledge. Transcendental arguments about the objects of experience therefore always depend on at least one empirical premise. We should thus not assume that synthetic a priori propositions are entirely independent of empirical investigation. Nor are they intended to counter an *absolute* sceptic. Rather, critical metaphysics aims to distil the mental preconditions common to all of us in the formation of experience. In this sense, our experiences acquire a somewhat objective status, as they are shared among those who possess the relevant cognitive faculties described by synthetic a priori propositions (Ewing, 1938, ch. 2). This does not, however, imply that such metaphysical synthetic a priori propositions are not truly a priori; they are, because these conditions are innate to any cogniser (Machina, 1972).

Kant clearly distinguishes transcendental metaphysics from the work conducted in the empirical sciences. This distinction is coherent from Kant's point of view, since empirical science, in his understanding, presupposes the notion of an 'object' when investigating objects through experimentation. The empirical sciences therefore cannot provide insights into the preconditions of objects, and thus offer no access to their metaphysical nature. Neuroscience, for example, investigates objects such as neurones and brains. But if we are seeking the conditions for experiencing objects, then a theory that itself depends on objects will not suffice.

However, it appears that Kant may have been mistaken on this point. While Kant derives metaphysical claims by means of transcendental arguments, it has been convincingly shown that critical metaphysics need not rely solely on first-person reflection from the armchair. As Louise Antony (2004) explains, an investigation into the a priori can be naturalised by means of scientific inquiry into (human) cognition. She suggests that '[t]he a priori elements of knowledge are those elements that reflect or derive solely from features of our native cognitive machinery' (p. 5). In other words, we humans possess an in-built cognitive architecture that enables us to perceive and interpret the world. This machinery may make implicit assumptions about reality, which are then projected into our experiences. These assumptions, being innate, can be investigated by empirical means and may reveal the underlying structure of the a priori, as they persist throughout all our reflective processes, independent of specific experiences.

Innate cognitive processes may give rise to certain beliefs, and these beliefs are justified by the reliability of the processes that produce them. The a priori, then, can be characterised as the conjunction of innateness and reliability: the former explains how we come to have a priori knowledge, while the latter accounts for its justification. These cognitive structures can be examined either through transcendental arguments or through more empirical approaches. Noam Chomsky's (1967) work on universal grammar is a prime example of the latter, even though credible alternatives to Chomsky's theory now exist. His project aimed to uncover the linguistic rules innately encoded in the human mind. Since the eighteenth century, the brain sciences have developed a mathematical language that characterises cognitive processes in abstract, algorithmic terms. These descriptions of mental processes do not necessarily presuppose or rely on particular empirical objects. The innate structures that underlie the a priori are thus expressible in algorithmic terms that describe how our in-built cognitive machinery processes data. The details of this will be discussed in Part II of this book.

If we now interpret Kant's notion of the a priori along these lines, his approach to metaphysics can be recast as a naturalistic endeavour. To reveal the metaphysical nature of objects, one first identifies innate, reliable cognitive processes and describes them in algorithmic terms. From these cognitive processes, one can infer the essential qualities of experienced objects. This approach is what I call the *Cognitive Metaphysics* approach. The term 'Cognitive Metaphysics' has been introduced in a similar spirit by Lieven Decock (2018), who suggests that we should turn to the cognitive sciences to gain metaphysical insight into 'the basic categories in the human mind that structure the representation of the environment' (Decock, 2018, p. 2). For a unified metaphysical understanding of ordinary objects, I propose that the cognitive sciences can indeed provide new metaphysical insights—once we adopt the Kantian metametaphysical framework outlined in this chapter, according to which metaphysics is entangled with epistemology.

Summary
According to realism, the essential structures of the objects of our thoughts are isomorphic to the structures of the world, since natural selection has shaped us to access the true structures of reality. This view is undermined by the fact that evolutionary success does not require veridical representations. These sceptical concerns about representation motivate a Kantian approach to metaphysics, one

that investigates the necessary conditions of experience in order to derive universal principles about the objects of experience. Cognitive Metaphysics naturalises Kant's approach by turning to the brain sciences to uncover the fundamental principles that reliably shape our cognition and determine the qualities of the objects in the manifest image.

Part II
The Predictive Brain

Chapter 9
The Problem of Underdetermination

Having outlined in the first part of this book a methodology for pursuing a naturalist metaphysics aimed at developing a general theory of reality, we now turn to applying this framework through the lens of contemporary computational theories of the brain. To do so, we must first gain a deeper understanding of how the brain functions. In this part, I not only introduce key concepts from modern brain sciences, but also trace the historical development of computational models of the brain and examine how they have been influenced by the philosophical debate between realism and idealism.

9.1 The Computing Brain

René Descartes (1641) famously stated that non-human animals are machines, lacking an inner mental life, and offered explanations of how certain psychological processes such as memory and perception might be realised in a mechanical system. He did not, however, believe that human thoughts could per se be realised in a material world. Instead, he postulated that they occupied a distinct ontological domain. In Descartes' succession, and with the advent of the Industrial revolution, it did not take long for broader circles of intellectual life to conceive of humans as mechanistic wholes, including their mental processes. As the eighteenth century French physician and philosopher Julien de La Mettrie writes: 'the body is but a watch ... man is but a collection of springs which wind each other up' (de La Mettrie, 1748, p. 9).

This is the kind of materialism we already find in Thomas Hobbes, who believed that the mind could be fully explained in mechanistic terms and described mental activity as tiny motions of matter that enable human reasoning. But how is it then possible that a world of matter allows for beings that have an inner life of their own? The clash haunts us once again: we think of ourselves as conscious beings with a

A. C. Schwaninger, *Cognitive Metaphysics*, Synthese Library 529,
https://doi.org/10.1007/978-3-032-05420-3_9

capacity to act as we please, and yet we are material hunks of matter that follow the laws of physics like leaves following the river's flow.

Resolving the clash requires us to revisit ideas that can be traced back to the seventeenth century philosopher Gottfried W. Leibniz and that find their culmination in our modern understanding of computers. Leibniz believed that thought could be decomposed into a mathematically precise language by means of which we could resolve intellectual disputes. He envisioned a logical calculus in which every conceivable problem could be expressed using logical symbols and manipulated according to a set of rules. This approach implies that any arbitrarily complex idea or argument could be broken down into simpler components and systematically analysed, much like solving a mathematical equation. By applying this formal language, Leibniz believed that intellectual disagreements could be resolved with the same clarity and certainty as mathematical problems, leading to more objective and definitive conclusions in philosophical and scientific debates.

This idea was later complemented by George Boole, who developed the formal apparatus that allows us to express the logical relationships between propositions of arbitrary complexity. His algebraic system introduced a mathematical framework in which logical statements could be represented using binary variables—true or false, or 1 and 0—thus making it possible to bridge the gap between abstract thought and computations performed by material objects.[1]

Both the idea that mental activity can be decomposed into simple steps within a formal language and the idea that these steps can be functionally realised by physical units culminate in what is known today as the *computational theory of mind*.[2] The theory holds that complex thought arises from many tiny thought units that are physically realised by a system in motion. This system performs operations on symbols via hard-wired processing schemes, as we find them in digital computers. The computational theory of mind suggests that intelligent thought can emerge from mindless matter in the same way that sophisticated software runs on computer hardware. Thus, if we understand how computer hardware relates to its software, we may also understand how the brain relates to the mind.

A digital computer uses electrical circuits to functionally encode the basic units of Boole's logical formalism, allowing it to manipulate symbols or propositions according to a given code or set of rules. This can be illustrated by showing how a very basic hardware device can encode the conjunction of two simple propositions, thereby encoding for a marginally more complex one.

Suppose we have a proposition P, such as *'x is purple'*. Its truth value can be represented symbolically as 1 or 0 (true or false), and physically realised by the presence or absence of an electrical current in a copper wire. Similarly, we can represent the truth value of a second proposition, $Q =$ *'x is a house'*, by a second wire. If electricity flows through the first wire, then P is true; if it flows through the

[1] For a more complete historical picture and the philosophical commitments of the 'mechanical mind', see Crane (2015).

[2] The assumptions made by the computational theory of mind is discussed in detail by Isaac (2018).

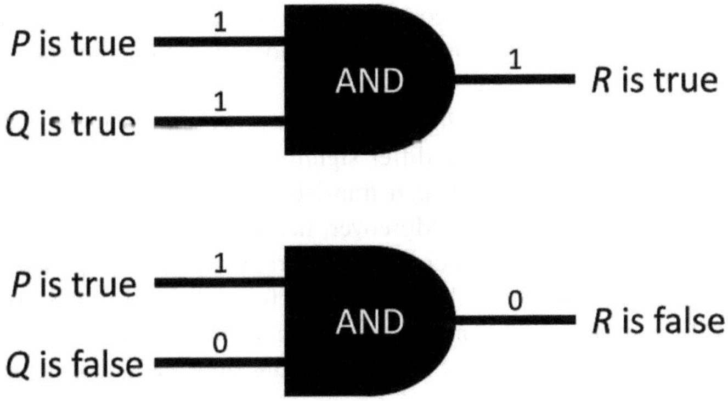

Fig. 9.1 How physical circuits realise logical conjunction

second wire, then Q is true. On this basis, we would like to physically encode a third proposition R = 'x is a purple house', which is equivalent to saying that both P and Q are true (formally: $P \bigwedge Q$). To do so, we need a wire that carries an electrical current flowing through it only when both P and Q are true, and thus only when there is electricity flowing through both of the initial wires.

To realise this, we introduce a physical circuit device built from electrical transistors, known as an 'AND gate'. This device allows current to pass through only when both input wires have a positive voltage. As illustrated in Fig. 9.1, the upper part shows both inputs active, resulting in a positive output current. In the lower part, the absence of current in the lower input (representing Q as false) results in no output signal, so R is also false.

The example illustrates how a machine can encode a more complex logical proposition by combining simpler ones. When electrical signals represent the truth values of basic propositions, a circuit like the one above can process them in a way that mirrors logical operations. In this case, the AND-gate ensures that an output signal occurs only when both input propositions are true, thereby physically realising the truth table of a Boolean conjunction. This process demonstrates how logical relations can be materially instantiated and transformed into new informational states. The output signal condenses the input information: it no longer indicates which of the two propositions is false ($\neg P \bigwedge Q$ or $P \bigwedge \neg Q$ or $\neg P \bigwedge \neg Q$) and instead relates information of different sources with each other that can be useful for further computation.

Abstractly speaking, information processing can be understood as the transformation of one set of possible states, each associated with a certain probability, into another set of states with different probability distributions. In the present case, there are four possible initial states ($\neg P \wedge Q$, $P \wedge \neg Q$, $\neg P \wedge \neg Q$, and $P \wedge Q$) and two possible output states (R and $\neg R$). Each of these states carries its own likelihood of occurrence. While there may be many purple objects and many houses in the world, the conjunction 'purple house' represents a rare event. As a result, when applied to arbitrary objects the circuit will rarely generate a positive output signal.

This means that the informational content of the output is typically low. This is not a flaw but a reflection of the system's purpose: to identify an uncommon and specific configuration.

While both the brain and the digital computer process information, the physical mechanisms by which they do so differ significantly. In digital computers, information processing is realised through transistor circuits, whereas in the brain it is carried out by biological cells. Moreover, the underlying hardware architecture diverges fundamentally. A digital computer clearly separates components responsible for data storage from those dedicated to performing logical and mathematical operations. In contrast, the brain does not make such a distinction: its processing units, the neurones, are involved in both storing and manipulating information simultaneously.

According to the computational theory of mind, the precise physiological underpinning of the brain is not essential for understanding it as a computational system. What matters is that symbol manipulation occurs according to rules that can be expressed in abstract, algorithmic terms. That is, in the language of computation. Thought, on this view, is an activity that can be described computationally. It arises from material processes governed by a sequence of rules that are continuously executed. The formal rules of computation can be specified independently of the system's physical substrate. Whether the brain is composed of biological tissue or silicon is irrelevant, so long as it implements the same computational rules.

By positing an analogy between mental and computational states, the computational theory of mind offers a remedy to the clash between the materialistic view and abstract entities like thoughts and ideas. It bridges the gap by suggesting that thoughts are the result of symbolic manipulations governed by rules, much like software running on hardware. The theory abstracts away from the physical specifics of the brain, focusing instead on the functional organisation of cognitive processes and implying that mental phenomena can be fully explained in terms of the brain's computational operations. This includes not only rational thought but also the entire human universe of emotions, artistic creativity and imaginative experience, all of which are seen as being encoded in a procedural, deterministic language of computation.

Despite the historical controversies, the computational theory of mind is today oftentimes perceived as radical only by novices. Many neuroscientists have not only accepted but also fully integrated the theory as the basic assumption of their approach for understanding the brain. Since the theory holds that the brain encodes and processes information in such a way that it can be described in computational terms, the real-world consequence is simply to treat the brain like any other dynamic system when modelling it mathematically. Although the notions of computation and information processing are useful for conceptualising what the mind does, all material systems encode information, and many can be said to process it as well. The growth rings of a tree can be interpreted as a symbolic representation of its age, and the smoothness of stones along riverbanks tells the story of how water has flowed over time.[3] Physical events leave traces in physical bodies, and these traces can be interpreted as representations of those events. Two such traces, originating from separate causes, can be linked and interpreted as a form of information processing, just as in the case of the AND-gate discussed above. For instance, the water concentration in a flower might symbolise rainfall levels (P = 'There is a good amount of rain'), while the nutrient concentration might reflect soil quality (S = 'The soil contains a good amount of minerals'). Together, the flower's health and growth can be seen as a symbolic representation of both favourable rainfall and soil conditions ($R \wedge S$), functioning analogously to an AND-gate. Information processing is omnipresent in the natural world and not unique to the brain. What is unique, however, is the complexity with which this processing occurs.

9.2 Discovering the Neural Code

According to the computational theory of mind, the processes of the brain can be abstracted as computational models specifying rules in algorithmic and mathematical terms that can, in principle, be run on a computer. Understanding the mind in these terms only appears counterintuitive because of the judgements we form from within the manifest image. It is therefore no coincidence that the most famous counterargument to the computational theory of mind, the so-called 'Chinese Room' argument, is largely driven by ordinary, everyday intuitions about intelligence. The argument suggests that there is a gap between truly understanding the meaning (semantics) of a statement and merely manipulating its symbols according to formal rules (syntax), as a computer does.[4] From a scientific point of view, however, the computational theory of mind simply makes clear that the brain is not a magical device lying beyond the ordinary physical realm. Describing phenomena in computational or mathematical terms is precisely what all scientific disciplines do when modelling dynamic systems. Scientists use computers to simulate the behaviour of physical processes at any scale: from molecular interactions to fluid dynamics to social systems. Importantly, even quantum mechanical processes can be simulated with high accuracy. Although simulations may fail to replicate experimental data due to our ignorance of certain causal connections or the effects

[3] Putnam (1988, pp. 121–125) and Searle (1990) have also criticised the computational theory of mind for being, in some sense, trivial for similar reasons.

[4] Searle's (1980) Chinese Room argument was intended to show that computational systems cannot generate genuine understanding. While it is widely agreed that Searle's argument fails, there is a lesson to be learned about the concept of intelligence: intelligent thought is typically attributed only to systems that process information rapidly. Searle's thought experiment highlights that a slow system does not seem intelligent. The person in the thought experiment who carries out symbolic operations by hand would require centuries to compute what a digital system could process in milliseconds. As a result, the room is intuitively denied the status of being intelligent. This link between speed and intelligence is evident in everyday expressions like 'he is a slow thinker'.

of chaos, there is no indication that we do not live in a computational universe, that is, one in which all processes could, in principle, be described in computational terms.[5] Since the brain is a material object, there is no good reason to believe that its processes cannot also be described in computational terms. The computational theory of mind can thus be understood as simply stating that scientists should not treat the brain differently from other objects in the material world.

These theoretical insights into how we ought to think about the mind were, at the beginnings of brain science, of lesser importance. Traditional neuroscience research was much more focused on understanding the physiology of the brain than on the computations it performs. The first pioneering work in this field comes from Santiago Ramón y Cajal, who discovered that the tissue in the neocortex consists of individual units called *neurones*, which communicate with each other across synapses and are electrically polarised (Ramón y Cajal, 1911). Since then, we have learned a great deal about the anatomical nature of these cells, including how they generate and propagate electrical signals, how their dendrites and axons facilitate signal transmission, and how synaptic connections enable intricate networks of communication within the brain. Advances in microscopy and molecular biology have further revealed the complexity of neuronal structures and the dynamic processes involved in synaptic plasticity, neurotransmitter release, and receptor activation.

However, while our understanding of the brain's physiology helps us better grasp the 'hardware' of the mind, a complete account of the brain and human cognition requires us to supplement these insights with an understanding of the computations and information processing taking place. If we accept the computational theory of mind, it makes sense to ask the fundamental question: What is the code of the brain? Or, in other words: What computations does the brain perform in order to process sensory information, make decisions, trigger behaviour, and generate conscious experience? We refer to these computations as the *neural code*.

Investigating the behaviour of neurones provides crucial insights into the neural code. Building on such insights, the discipline of *Computational Neuroscience* (also sometimes referred to as Theoretical Neuroscience) is dedicated to uncovering the neural code on the basis of neuroscientific findings. Computational neuroscientists develop computer models that aim to capture how the brain processes information and compare the predictions of these models with empirical data.[6]

[5] It has been argued that qualia are not part of the physical, mechanical world or at least cannot be described in such terms (see, e.g., Chalmers, 1996). However we might think about qualia, this will have no bearing on the further discussion about material objects.

[6] Some philosophers, such as Searle (1980), and scientists, such as Penrose (1989), have rejected the idea that the brain performs operations describable in computational terms. Penrose, for example, suggests that certain quantum mechanical effects in the brain might introduce a non-computational dimension to cognition. However, since quantum effects can also be described computationally, and given that it is unlikely that relatively large structures such as neurones are affected by non-computable phenomena at the quantum scale, this view is not widely accepted among scientists.

One might wonder whether the project of discovering the neural code is ultimately futile, given the vast number of regions and cell types in the brain, all of which may function simultaneously using different neural structures. From this perspective, it may seem impossible to achieve the goal envisioned by computational neuroscientists. This worry is, however, unjustified given what we know about the brain. As far as scientists can tell, the neocortex is one of the largest parts of the brain and is responsible for processing sensory perception, cognition, language, and coordinating body movements. This region has been intensively studied and has proven to be unexpectedly uniform in structure (Hubel & Wiesel, 1974; Rockel et al., 1980). For this reason, the reasonable assumption has been made that a fundamental neural circuit repeats itself throughout each cortical area (Creutzfeldt, 1977; Szentágothai, 1978). This implies that, at least for the computational components essential to how structural features in our perception arise, we can make significant progress by studying smaller regions of the brain and generalising the discovered principles to other cognitive functions.

But if the brain is so homogeneous, how then can it give rise to the complex thoughts we humans are capable of? What makes the brain unique is not, as Descartes believed, its capacity to connect to an immaterial soul, but rather its sheer complexity, stemming from the number of parameters it controls, far exceeding those of any other known natural system of comparable size. Complex systems in the biological world are typically characterised as lying somewhere between the rigid order of a crystal and the chaotic randomness of a soup of independent particles (Huberman & Hogg, 1986) (see Fig. 9.2). They exhibit properties that are difficult (or maybe even impossible) to reduce to their parts and involve interacting subsystems.

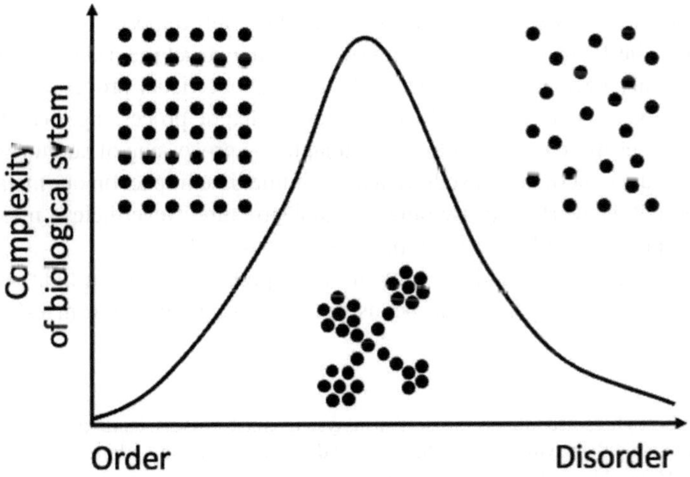

Fig. 9.2 Variation of a system's complexity with its degree of order. A complex system in the biological world is best described as a system that is between the two extremes of order and disorder. Image inspired by Huberman and Hogg (1986)

In an information-theoretic sense, biological complex systems are not maximally complex, since maximum complexity is associated with pure randomness. Random systems require the longest possible description length and cannot be compressed, thereby carrying the highest information content (as discussed earlier in connection with real patterns). However, this notion of complexity is of limited relevance in the biological context. A system qualifies as biological only if it maintains structural integrity over time, and complete disorder does not allow for such stability or organisation.[7] The brain, in this respect, is a highly complex system: it is capable of sustaining itself and the rest of the body while at the same time controlling an unimaginable large number of different possible states.

The most fundamental unit for describing the computational state of the brain is the neurone with its synaptic weights, and their connections to other neurones.[8] The brain contains somewhere between 80 and 100 billion neurones, and an estimated 100 trillion synapses through which neurones connect and modulate the strength of these connections.[9] Each neurone can be in an active or inactive state, sending signals to other neurones via its synapses. Each synapse can vary in strength, influencing how actively it transmits signals, and synaptic connections can be removed or reconfigured with other neurones. Combined with mechanisms for transitioning between these states, this gives rise to an extremely rich representational structure and computational capacity that far exceeds anything else known in the universe.

9.3 Underdetermination and the Inverse Problem

Now that we understand in what way computation is a useful notion for analysing the brain, and that we can sensibly talk about the brain's *neural code*, we can begin to consider how this relates to the structures the brain imposes or fails to impose on reality. The focus here will be on how the brain structures *visual* sensory data. This is mainly because the number of theories about visual processes throughout history far exceeds those available for auditory or haptic processing. In addition, the brain dedicates more resources to vision than to the processing of auditory or tactile signals. Visual experiences also have a more immediate impact on our metaphysical conception of the world as possessing essential structure. Nevertheless, most of what I present applies equally to other types of sensory signals.

Let us therefore take a closer look at visual processing. When light reflects off matter and enters the eye, the lens bends and focuses the light to form a sharp image

[7] Some philosophers accordingly argue that a system must possess representational properties in order to manipulate symbols, and that this condition is not met by arbitrary physical systems (Fodor, 1998, pp. 11–12; Ladyman, 2009; Sprevak, 2010).

[8] Some have, however, hypothesised that other components in the brain, such as glial cells, might also have computational significance (De Pittà & Berry, 2019).

[9] These number estimates vary greatly in the literature; one estimation is given by Stix (2013).

0.8	0.8	5.0	5.0	5.0
0.0	2.7	3.3	5.0	5.0
1.8	3.7	1.7	0.0	5.0
1.0	0.0	1.6	4.1	5.0
5.0	1.4	3.3	3.9	5.0
5.0	5.0	2.5	4.7	3.7
5.0	4.5	1.7	3.9	4.5

Firing rate

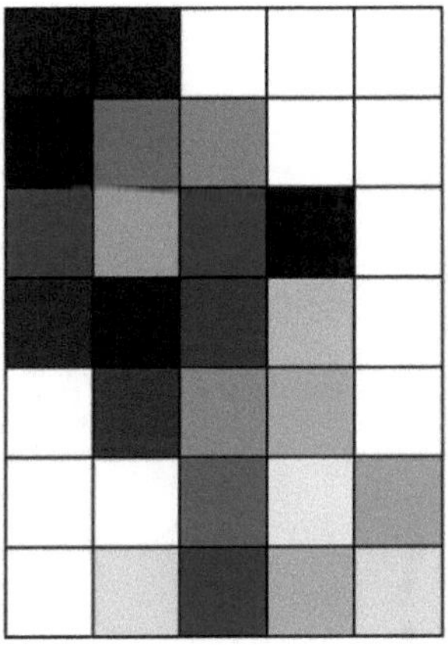

Colour map

Fig. 9.3 A fraction of the visual field displayed in terms of cell activation (e.g., firing rates of the postsynaptic neurones that are activated by potential shifts of the photoreceptor cells)

on the retina. The retina contains over a hundred million photoreceptor cells, which become activated by light and relay the incoming information deeper into the brain. Since the brain can, for our purposes, be understood as a computing machine that performs information processing, we may treat this sensory input simply as input *data*. The complete set of data received by the retina is referred to as the *visual field*.

The data available to the brain is always discrete and finite, represented by a finite expression. The simplest reason for this is that neurones are either active or inactive, and therefore can propagate only finite, rather than continuous or infinite, signals. The best way to conceptualise visual sensory data is to think of it as a matrix of values arranged on a two-dimensional grid, similar to the pixels visible when zooming in on a digital image. Each value in this grid corresponds to the activity of a neurone, which is triggered by the electric potential shifts of its connected photoreceptor in the receptive field. Like the pixel values in a camera, these values can be interpreted as colours within an image. This is illustrated in Fig. 9.4, where a small section of the visual field is displayed both in terms of photoreceptor activity and the corresponding colour interpretation. What is important for our purposes is that the raw data received by the retina contains no explicit structural information. The image offers no indication of boundaries, nor does it reveal whether we are looking at independently structured objects or at the structureless forms of a Jackson Pollock painting. The visual field itself lacks structure, and from the image alone, it is not at all clear where objects begin or end.

According to the realist, the task of the brain is to *infer* from the provided pixelated data, together with certain additional assumptions to which the brain has adapted over the course of evolution, what the structures in the image are and what

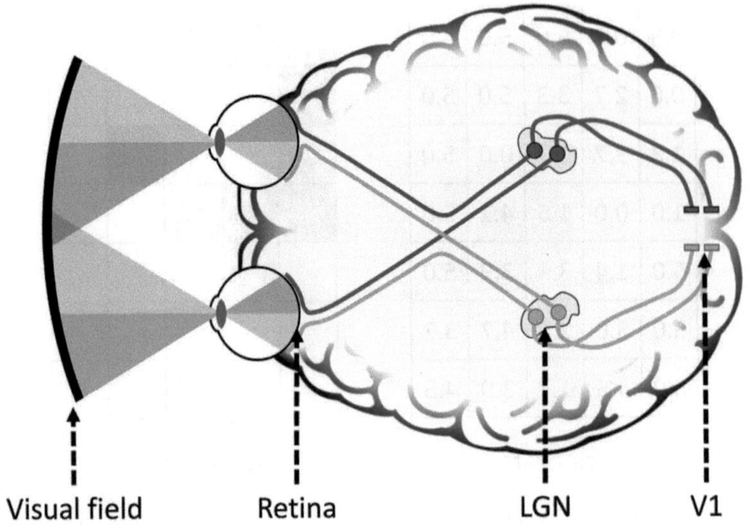

Fig. 9.4 Visual pathway of sensory information from the eye to the visual cortex

we should expect to find *out there* in the world. The idealist, on the other hand, will argue that there is neither structure in the data nor in the world itself, and that it is therefore the brain which must impose structure onto the data. To reveal the environment's underlying structure (as the realist would say) or to *impose* structure onto the data (as the idealist would prefer to say), the sensory data is projected along the visual pathway into different regions of the brain, as illustrated in Fig. 9.4. The eye serves as the entry point into this visual pathway. The right half of the visual field is processed by the left hemisphere of the brain, while the left half is processed by the right hemisphere. Both streams of visual information are projected to the lateral geniculate nucleus (LGN), a key structure in the brain's thalamus, from where they are further relayed to the primary visual cortex (V1).

The unstructured nature of sensory data arises from the universal fact that the stimulations received by the senses of any cognitive system are underdetermined by their causes. The visual data contains insufficient information about how the world is structured, and yet the brain must still succeed in generating the experiences we have. As the vision scientist Richard Gregory explains, in terms that lean more towards idealism than realism: 'Perceptions are *constructed*, by complex brain processes, from fleeting fragmentary scraps of data signalled by the senses' (Gregory, 1972, p. 2, emphasis added). The neural code must therefore offer a solution to how these scraps of data are transformed into our rich visual ontology, which corresponds to the structured world of the manifest image, with its coherent objects, boundaries, surfaces, and motions. This challenge is known as the *problem of underdetermination*, and it must be addressed by any cognitive system.

Any cognitive system must turn the given sensory data into a structured format that allows the system to reason about and conceptualise the world. Since the essential structural aspects that allow for such reasoning (such as the essential structures of objects) are not entailed in the data, the information about the shapes of objects and their relations to one another, as well as their parts, 'must be derived from the sum of retinal information together with various assumptions about the structure of the world' (Shapiro, 2011, p. 29). These assumptions about the world are not necessarily veridical or indicative of primary properties, as a realist might suggest, but could, as the idealist would argue, merely enhance the fitness of the cognising system without being isomorphic to the way the world actually is.

The realist interpretation of the underdetermination problem holds that any cognising system is required to find a way to invert the mapping from external causes to sensory effects. This is a difficult task, given that different causes can result in the same effects. The realist calls this the *inverse problem of perception*, as it involves deriving the ontology of the world from our experienced ontology. She assumes that there is an isomorphic relationship between mind and world, and that this relationship can, by definition, be inverted. For the idealist, such an inversion is not possible, at least not for any essential structural aspects, because the structures we perceive in the world are imposed by us. The problem is therefore not one of inverting a given structure, but of explaining how structure is constructed.

In contemporary brain science literature, the inverse problem and the problem of underdetermination are often treated as virtually synonymous, largely because realism is the default assumption for most researchers. It is important to recognise, however, that referring to the inverse problem already involves realist presuppositions, which, as argued throughout this book, are mistaken.

9.4 A Multiplicity of Possible Experiences

The significance of the underdetermination problem is not intuitive. When we look at the world, we have the impression that we are effortlessly and passively perceiving what is in front of us. Do we not, just by looking, know what the objects around us are like? It seems obvious that our higher cognitive abilities allow us to detect objects in the sensory data without having to rely on any lower-level, unconscious operations. While such lower-level computations support the process of object detection and make consciousness possible, it may appear that once a being is conscious and capable of accurately interpreting visual data, it should also be able to detect structures in the world. After all, the cognitive capacities required for consciousness and intelligence seem to entail the ability to identify something as basic as structure in visual sensory input.

But the intuition that consciousness entails the ability to identify structures of the world in visual data is mistaken. This is illustrated by rare cases in which the visual area of a person's brain has been damaged due to a stroke or an accident, leading to abnormal object perception, despite the patient being fully conscious and having no deficiencies in receiving or projecting visual data to the brain, as would be the case with an eye impairment. There are, for instance, clinical patients who

> can see objects change their positions but cannot see them move—a syndrome that a philosopher once tried to convince me was logically impossible! The stream from a teapot does not flow but looks like an icicle; the cup does not gradually fill with tea but is empty and then suddenly full. (Pinker, 1997, p. 19)

Such a medical condition shows that the initial intuition is incorrect. Even though the patient is fully conscious, he or she experiences the world very differently, despite having access to the same sensory input as any other person. While this patient does perceive the world to be Px-structured, other structural aspects in their visual ontology do not match with the *essential structures* any healthy person perceives when looking at an object. In a healthy visual ontology, the location of objects and their spatial and temporal relations can be described by a continuous function. This is structurally different from the world experienced by the patient. Whereas we normally perceive objects to move continuously through space and time, the patient in this case perceives them as behaving discontinuously—without lacking access to the relevant visual data.

Underdetermination has the peculiar consequence that, for a given cogniser receiving sensory data, there exists a *multiplicity of possibilities* regarding what the experienced qualities of ordinary objects could be. Under normal conditions, the information the brain imposes onto the sensory data enables us to experience structures in the familiar way. As with any biological process, however, it is possible that this mechanism of blending sensory input with our own neural structures may malfunction. The clinical case of a patient who perceives objects to change positions without seeing them move illustrates such a breakdown.

It is a remarkable fact that there have been clinical cases in which patients do not even perceive a Px-structured world, despite having access to the visual image. Patients who suffer from *visual agnosia* fail to see objects in the first place, even though they can see the sensory input. They are not blind and are capable of seeing coloured images in front of their eyes, yet they do not perceive any structure in these images.[10] People suffering from this rare condition report that they are no longer able to identify objects, let alone any of their proper parts, even though they perceive the world clearly. A person with this condition may look up into the sky and see blue, yet never see a *thing* that is blue, such as the sky. She may look at a tomato and see redness, yet will never be able to see the red patch of her experience as a property of a certain *thing*. All she perceives are disconnected patches of colour but no objects with properties.

The condition results from a lesion in a certain region of the primary visual cortex. It sometimes has little to no effect on the person's memory or abilities to reason, which is why individuals with this condition are able to report the unusual nature of their experiences. A patient may recall from before the brain damage

---

[10] This is possibly due to the absence of certain predictive circuitry or a failure to predict sensory input accurately, which, as we will see later, is a crucial element of experiencing the Px-structure. Further details on these disorders and how they relate to modern views in cognitive science are discussed in Ptak et al. (2020).

that the blue expanse she sees above her is the sky, yet she will insist that she no longer sees the sky as a *thing* that is distinguishable from other things. The visual information received by the retina lacks usable structure and is transmitted in this unstructured form to the patient's conscious perception. As a result, the information offers almost no practical assistance in everyday life, aside from supporting very general inferences based on the typical colours found in certain environments. She may distinguish between being in a forest, where greens, browns, and blues dominate, and being inside a building, where whites and greys prevail. She can tell the difference between day and night, or between dark and bright rooms. Beyond that, however, she behaves much like someone who is completely blind, as perceiving a structureless world does not enable one to effectively navigate through it.

Visual agnosia illustrates that a Px-structured visual ontology cannot be taken for granted, even in the case of a conscious being receiving ordinary sensory input. It also reveals that our notions of *seeing* and *perceiving* can be ambiguous. When someone suffering from visual agnosia looks up at the sky, she does not see the sky in the same sense that we do. Although she is clearly seeing, she is not seeing some-*thing*. She perceives no objects, even though the sensory data reaching her brain is identical to that of a healthy person. This is what can be called *unstructured* seeing.

The condition is not akin to a perceptual impairment such as strong astigmatism, which interferes with the accurate collection of sensory data and leads to a blurred perception of objects. In the case of visual agnosia, the information reaching the senses is intact, but the brain fails to impose the necessary structure onto it. According to the realist, this structure corresponds to how the world truly is. According to the idealist, it is a neural construct that creates the illusion of a reality populated by independent individuals organised according to a Px-structure. In either case, one may refer to this as *structured* seeing.[11]

Is there any way we can experience unstructured seeing ourselves? Have a look at Fig. 9.5. At first glance, there is nothing recognisable in the image. It appears as a collection of black and white patches arranged in seemingly unstructured fashion. Very quickly, however, our brain begins to group certain areas together. The circular black area in the top left becomes conceived as a single unit. With some effort, we eventually shift to a structured mode of perception, recognising the Dalmatian dog in the centre of the image, sniffing the ground and walking away from the photographer toward a tree. Once the dog is detected, it is typically impossible to see the image in the same way as before, as an assemblage of disorganised black spots. The contrast between our first glance and the perception that emerges once cognition has imposed a certain structure onto the image roughly illustrates what unstructured seeing might feel like in contrast to structured seeing.

[11] Related distinctions, though developed from machine learning contexts rather than from clinical observations, are discussed in Smith (1996).

Fig. 9.5 Image from Gregory (1970)

9.5 Gibson's Theory of Perception

In line with the widely accepted realist worldview, most philosophers and cognitive scientists describe the problem of underdetermination in terms of the inverse problem. This implies that the brain has the capacity to reconstruct the 'properties of external objects' (Wiese & Metzinger, 2017, p. 4), although much of our visual ontology—that is, the ontology of our experiences—is not necessarily a reconstruction of how the world actually is. The problem, as formulated within this realist framework, presupposes that the world consists of objects structured in a pre-given Px-structure, and that science must explain how the brain retrieves information about these objects and their properties from sensory data. In this formulation of the problem, the Px-structure is taken to exist mind-independently.

Classical approaches to the underdetermination problem are realist in that they adopt a minimal set of assumptions about the world or even embed structured objects directly into their theoretical frameworks. An instructive example is James J. Gibson's theory of perception (1966), which addresses neural computation by focusing solely on the relationship between observer and object, rather than on internal brain processes. Gibson defines an object in realist terms as a 'persisting substance with a closed or nearly closed surface and [that] can be either detached or attached. . . . [T]he surface of an object has a definite texture, reflectance, color, and layout' (Gibson, 1979, p. 34). For Gibson, the perceived 'structure depends on sources in the outer environment' (1966, p. 267).

So how then, according to Gibson, does the brain overcome the problem of underdetermination? He argues that sensory information should not be understood as a single static image, but rather as a sequence of visual inputs that stand in relation to the subject's position and movement through the world. This dynamic

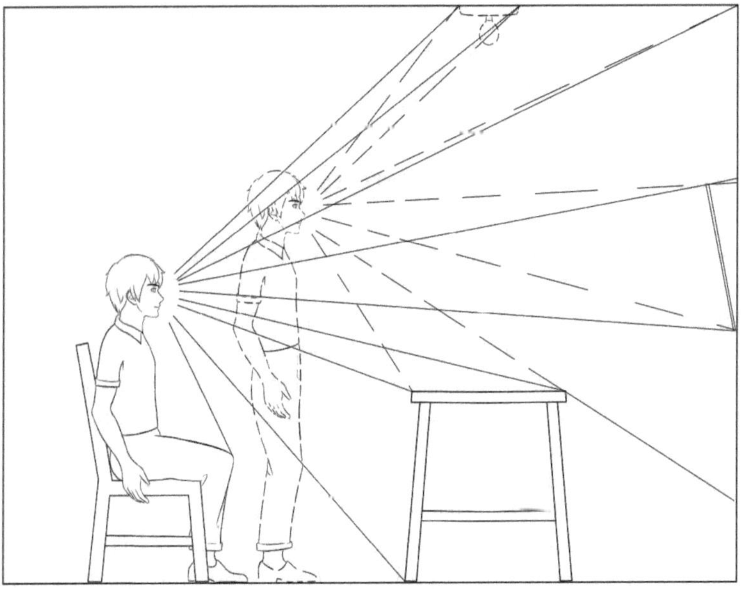

Fig. 9.6 Optical arrays change by the subject's movements. Adapted from Gibson (1969, p. 196)

relationship provides sufficient information to accurately reconstruct all the objects surrounding us. His central idea is that when we move from one location to another, the optical arrays change accordingly. Since the brain also registers how the body has moved, it can apply triangulation to determine object depth and spatial relations. This idea is illustrated in Fig. 9.6, which shows the shifting light arrays as a man stands up, allowing the brain, according to Gibson, to infer the structure of the environment (Gibson, 1969).

Gibson's theory not only provides an example of how the problem of underdetermination can be resolved from a realist perspective, it also demonstrates that although the neural code can be described without reference to internal brain processes, our understanding of it may nonetheless depend on relating the world to the mind. His theory further suggests that bodily movements can be coupled to perceptual experience. For the purposes of this discussion, however, Gibson's account serves primarily to illustrate a case in which a mind-independently structured environment is assumed and correlated with the relative position of the observer in order to explain how the brain might resolve the problem of underdetermination.[12] Gibson's approach is remarkable in that it investigates the

[12] While Gibson is a realist in the sense discussed here, he does acknowledge a subjectivist element in his view. He notes that locomotion varies across species and, as a result, so does visual ontology. Nevertheless, his theory treats the basic Px-structure as universal, and he does not assign any substantive subject-relative role to this aspect of perception.

neural code without requiring detailed knowledge of anatomical processes inside the human brain. Most theories in neuroscience, however, focus on the internal structures and mechanisms of the brain. Let us now turn to these approaches.

9.6 Hard-Wired Feature Detectors

Despite very limited neurological findings, simple computational models of neural processing had already been developed in the first half of the twentieth century to offer possible explanations of how the brain might computationally register the presence of a certain feature in the visual field. One influential model is the so-called *Reichardt detector* (Hassenstein & Reichardt, 1956), which was postulated as a hypothesis to explain how biological neural circuits can detect the direction and speed of a moving stimulus, such as a square-shaped object traversing the visual field from left to right (Fig. 9.7, top).

The model rightly assumes that the human retina consists of sensory receptors that jointly detect the luminance at their position. In the presence of a stimulus, a specialised group of receptors project their activation through a channel to a neurone, indicating the recorded presence of the stimulus. This is illustrated in Fig. 9.7, where at time t_1 the presence of a stimulus is detected by a group of receptors A. Moving at a velocity v from left to right, the stimulus activates at time t_2 another group of receptors, B, that are located in the retina at a distance Δx from A.

The model hypothesises that the activations of A and B each generate electrical signals that travel through separate channels to a downstream neurone, C. Crucially, C only activates if the signals from both A and B arrive *simultaneously*. The activation of either A or B alone is insufficient to bring C's membrane potential to the threshold required for firing an action potential.

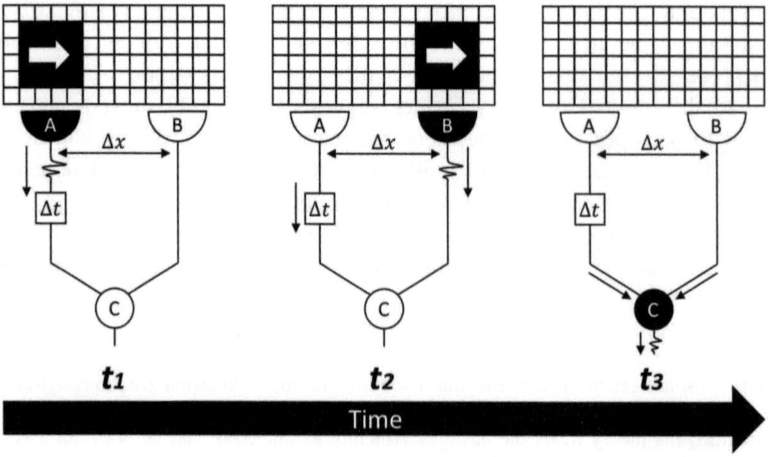

Fig. 9.7 Reichardt detector

If the channels from A to C and from B to C are of equal length, and assuming constant propagation speed for the electrical signals, then neurone C will never activate in response to a single moving stimulus. In that case, the signal from A would cause a temporary rise in membrane potential, which would decay before the delayed signal from B arrives. As a result, no action potential would be triggered.

Instead of using equally long wires between A–C and B–C, the wiring between A and C can be extended, causing an additional delay Δt for the electrical current to arrive at C. A moving stimulus at a velocity $v = \frac{\Delta x}{\Delta t}$ will then result in a consecutive activation of A and B, each generating a signal that reaches neurone C at the same time t_3. In this case, neurone C activates and forwards an electrical signal that indicates the presence of a stimulus moving from left to right at velocity v. The activity of neurone C therefore encodes the information that an object at a certain velocity has passed through the visual field. This information is then passed on to other locations in the brain.[13]

The Reichardt detector model was inspired by the analysis of the optomotor response in insects (Hassenstein & Reichardt, 1956) and was later interpreted as corresponding to the motion-detection machinery in humans (Reichardt, 1990). It is one of the earliest models of motion detection and suggests that there is an innate, hard-wired, immutable machinery in the brain for motion detection that enables it to draw certain conclusions about events in the environment.

The Reichardt detector is just one simple model of how neurones can be wired to compute the presence of specific features in the visual data, such as the motion of a stimulus. Similar models have been developed to detect other features. For instance, if a group of receptors are aligned along a straight line and are wired to a single neurone using equal wire lengths, then they can jointly detect an object with a straight edge that is present in the visual field at the appropriate location and angle. The same applies to all sorts of shapes, such as edges and circular curvatures. The general idea behind these models is that the brain possesses a highly complex feature-detection apparatus that identifies features in the visual sensory data and infers from this what objects must be present.

The confirmation that these models were on the right track came through the monumental work of David Hubel and Torsten Wiesel on the visual cortex, for which they received the Nobel Prize in 1981.[14] Their research revealed a hierarchical organisation of the visual cortex, suggesting that visual information is processed in a step-by-step manner. In their work, we again find the idea of pre-given objects and essential structures located in the world without our own contribution. Their understanding of how the brain relates to the world reflects the straightforward realist view: ‘images produced by an object fall on corresponding parts of the two retinas’ (Hubel & Wiesel, 1962, p. 149), presupposing that there is a plurality of objects *out there* for the brain to detect.

[13] Notice that neurone C can be interpreted to function like an AND gate that is active when two conditions are fulfilled simultaneously: the presence of a stimulus at the position of B and its presence a short moment earlier at the position of A.

[14] See, e.g., Hubel and Wiesel (1962, 1968).

The information from these images is transmitted from the retina to the visual cortex, where Hubel and Wiesel developed a model that would dominate vision science for the next fifty years. In their experiments, they presented cats with different visual stimuli and measured the activity of individual cells in the primary visual cortex. Using electrodes inserted into single neurones, they found that some neurones responded exclusively to the presence of specific features in an image, such as particular curvatures, angles, lines, or movements.[15]

In the presence of other stimuli, these cells showed no or very little activity. The feature-selective cells in the visual cortex are what Hubel and Wiesel termed *simple cells*, and they appeared to demonstrate that innate neural circuits are dedicated to detecting specific features. For the next two generations of scientists, the classical understanding of the visual system was defined by Hubel and Wiesel's theory: simple cells are hard-wired to detect highly specific features in the visual field, such as a horizontal edge moving from left to right.

9.7 Quine's Three Assumptions About the Brain

The results of Hubel and Wiesel were not confined to the neurosciences but quickly spread among academics, becoming general knowledge and influencing the thinking of many philosophers. Let us understand classical neuroscience through the eyes of Quine, who is not only of particular importance to metaphysical inquiry but is also often regarded as one of the most influential figures of twentieth-century analytic philosophy, shaping its general approach and consensus.

In an essay on how the sciences can ideally record their empirical findings and generalise them into theories, Quine makes numerous remarks on neuroscientific findings, which he believes support his convictions about the relationship between human cognition and the objects of scientific inquiry. In this context, he refers to Hubel and Wiesel's work to illustrate that

> ... overall traits of the scene are selected and communicated as wholes. One bank of brain cells responds *exclusively* to scenes in which there are some conspicuous diagonals from upper right to lower left. Other banks of cells *specialize in other broad traits*. Various traits thus abstracted get superimposed to kick off the appropriate response. (Quine, 1993, p. 114, emphasis added)

Thus, in accordance with what we have discussed above, Quine interprets Hubel and Wiesel as showing that feature detection is hard-wired in the brain. From the

[15] To the animals' detriment, many of these experiments required one eye to be fixated while the other was deprived of vision to limit sensory input. In some cases, one eye was entirely removed, and to prevent eye movements, the cats were paralysed with succinylcholine. In experiments on monkeys, the animals were trained to fixate an image on command. It was not uncommon for scientists at the time to suture the eyelids of one eye while mechanically holding the other using a clamp attached to a frame fixed around the animal's head (see, e.g., Blakemore et al., 1972). A dreadful act in the venerable quest for knowledge.

fact that the brain can identify certain features in a visual image, Quine elsewhere suggests that our innate feature-detection machinery enables us to identify ordinary objects:

> There is a harbinger of [the positing of objects] already in our *innate* propensity, and that of our animals, to confer salience on those components of a neural intake that transmit corporeal patches of the visual field. (Quine, 1995, p. 254, emphasis added)

Quine claims that our perception of similarity between things arises from prior experiences and interactions with them. He believes that the cognitive mechanism involved in conjoining the detected features into a single object-experience is based on making similarity comparisons between the detected features and previous experiences or learned associations, all of which are determined on the private, individual level:

> Neural intake is private, for subjects do not share receptors. Perceptual similarity, then, is a relation between a subject's neural intakes. Though testable, it is a private affair; the intakes are *his*, and are perceptually more or less similar for *him*. Perceptual similarity is the basis of all learning, all habit formation, all expectation by induction from past experience; for we are innately disposed to expect similar events to have sequels that are similar to each other. (Quine, 1995, p. 253)

According to this traditional way of thinking about the brain, the perceptual part is primarily concerned with extracting information from proximal stimulation in order to obtain accurate representations of the world. The innate feature detectors have, on this classical view, evolved to accurately detect what exists in the external world. Perception is therefore seen as largely dependent on universally shared mechanisms that detect specific features in the perceptive field. Once enough features have been detected, the system can infer what objects lie 'out there'. It can solve the problem of underdetermination by relying on prior experience to determine that certain feature combinations correspond to certain objects. For example, the detection of three connected edges is interpreted as a triangle. Two pairs of parallel edges that are perpendicular to one another are interpreted as a square. A square with a triangle on top is interpreted as a house.

A consequence of thinking that the brain possesses an innate feature detection machinery that responds in a universally shared and determinate manner to the features of a stimulus is that perception becomes characterised as a one-directional causal process. It assumes a dominant causal chain running from a physical object to its neural representation and, ultimately, to the utterance made by the speaker. Quine elaborates on this as follows:

> Causal continuity is the fact of the matter: the causal chain from Mama or the rabbit to utterance of the observation sentence 'Mama' or 'Lo, a rabbit'. Psychologists fix upon one or another point on this causal chain and call it the stimulus, to which the response is conditioned. (Quine, 1993, pp. 113–114)

In alignment with the classical neuroscience description, Quine thinks that light is reflected from an object in the world and falls onto our retina. From there, it activates certain cells in the visual cortex that interpret the scene as containing objects of a certain kind, leading to the activation of other regions in the brain that result in specific actions and utterances.

It is no wonder, then, that philosophers like Quine had little further interest in what the brain was doing and were drawn more to understanding people's utterances and vocalisation of thoughts rather than the cognitive underpinnings of these thoughts. If cognition were correctly characterised merely as this one-dimensional, linear process, there would be no genuine philosophical need to engage with it further, since everything relevant for understanding reality would already be externally present in the world.

In conclusion, Quine makes three assumptions about the brain based on the common neuroscientific knowledge of his time. First, we have an innate feature detection machinery. Second, objects are detected as sets of certain features based on comparing how similar these features are to our learned and established object categories. Third, Quine assumes cognition can be approximated as a one-directional causal process. From these assumptions, the realist position follows naturally, as the brain's task is to passively decode the information it receives and thereby reveal the true structures of the world. These assumptions, implicitly or explicitly, appear in various philosophical texts. For instance, the manner in which Ladyman and Ross discuss real patterns, and how they conceive these as individuals, can be traced back to Dennett's original paper, in which he writes:

> [A pattern is] discernible to the naked human eye ... because of the particular pattern-recognition machinery hard-wired in our visual systems – edge detectors, luminance detectors, and the like. (Dennett, 1991, p. 33)

All these philosophers give little relevance to the neural computations taking place and pursue a realist approach to material objects. We will see how modern neuroscience shows that each of Quine's assumptions is false and that there is little support for the realist position.

Summary

To a large extent, the functioning of the neocortex is dedicated to solving the problem of underdetermination. Hubel and Wiesel's discovery of simple cells in the visual cortex made plausible the idea that the brain has an innate, hard-wired machinery for detecting various features present in the visual field. Their findings led many philosophers to conceive of the neural code as a one-directional causal process in which the brain computes similarities within sensory input. This interpretation naturally aligns with realism, framing the underdetermination problem as an inverse problem of perception, whereby the brain *re*-constructs the structure of the external world.

Chapter 10
Discovering Grouping Principles

As illustrated by Quine's writings discussed in the previous chapter, the standard cognitive science approach assumes that perception operates as a one-directional causal process. Although this view has dominated for many decades, we will see in this chapter that neurological findings present a different perspective. Notably, some researchers as early as the 1910s had already begun to challenge this conventional view.

10.1 The Lateral Geniculate Nucleus

When information about the external world is received through our eyes, it is projected to a region of the thalamus called the lateral geniculate nucleus (LGN), from where it continues its path to the primary visual cortex (V1) for further processing. Traditional neuroscience research has dedicated most of its attention to computations occurring in the visual cortex (e.g., Hubel & Wiesel, 1962), supporting a realist worldview by positing a one-directional causal chain of neural processes involved in extracting structural information from sensory data.

However, significant neurological findings suggest that the brain plays a far more active role in processing information than previously believed. It is now established that only roughly 20 % of the information entering the LGN originates directly from the senses (see Fig. 10.1). The remaining 80 % comes from other parts of the brain, effectively convoluting the sensory information before it proceeds to the visual cortex (Churchland & Sejnowski, 1988; Varela et al., 1991, pp. 94–95). Given that only a small fraction of the information arriving in V1 is purely sensory, the initial assumption that V1 primarily performs basic processing to extract features exactly as they exist in the world appears far less plausible. In light of these findings, it becomes more reasonable to think that the brain does not merely extract objective

A. C. Schwaninger, *Cognitive Metaphysics*, Synthese Library 529,
https://doi.org/10.1007/978-3-032-05420-3_10

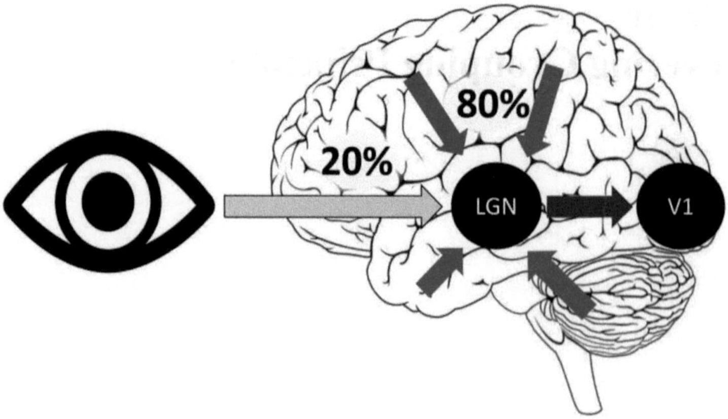

Fig. 10.1 Only 20% of the information received in V1 comes from visual stimulation; the remaining 80% originates from other parts of the brain

features from sensory data but rather actively *imposes* a substantial amount of internal information and subjective structure onto what it receives.

When I say that the brain is *imposing* information onto the sensory data stream, I mean that higher cortical regions, regions that the sensory information might never reach or only reach at a later stage of neural processing, add further information into the stream of data that is neither implicitly nor explicitly contained within the sensory input itself. The fact that higher regions in the brain impose information onto the sensory data stream challenges the standard view that cognition must be understood as a one-directional causal process. Instead, it suggests that while some information flows upstream' from the senses to higher cortical regions, other information simultaneously flows downstream', in the opposite direction, resulting in a two- or multi-directional process. Although the evidence presented here alone is not conclusive regarding multi-directional information processing, many other studies have arrived at similar conclusions (see, e.g., Gershman, 2019).

The computational models by Reichardt and the related neurological model by Hubel and Wiesel presuppose a one-directional computational flow under realist assumptions. In light of the above findings, these models and assumptions appear limited. But why, then, would the brain impose additional information on sensory data, and what would be the benefit of a two-directional process? To answer these questions, let us revisit some earlier approaches that were disregarded for most of the twentieth century but have gained renewed attention in recent years.

10.2 Gestalt Psychology

When sitting in a movie theatre, what we see on the screen are people and objects in motion. However, we all know that we are not *really* seeing movement; what we are actually seeing is a projection of individual images onto the screen that only *appear* to be in motion but actually the images themselves are all static. The human visual system can process 10 to 12 images per second and perceive them individually. However, once the image rate is increased, they are perceived as continuous motion. This effect inspired the scientist Max Wertheimer to formulate a radical hypothesis about the brain.

On a train ride from Vienna to Germany's Rhineland in 1910, Wertheimer looked out of the window and noticed how objects would disappear when occluded by other objects and then reappear due to the motion of the train. This creates a flickering effect familiar to anyone who has looked out of a moving train. The flickering of images led Wertheimer to reflect on the relationship between the visual stimulus and our experiential sensations, and he realised that the two are not always aligned. While the images we perceive flicker due to intermittent occlusion, we nonetheless experience a continuous background landscape.

The train ride inspired Wertheimer to a new theory that, in a Kantian spirit, traced experiences back to neural structure rather than to the stimulus alone. Excited by the idea, he disembarked early in Frankfurt, went to a toy store to buy a stroboscope—a device that flashes sequences of pictures at a rate that gives the impression of continuous motion—and experimented with it in his hotel room. The stroboscope demonstrated that we sometimes perceive motion where there is none, just like in the case of cinema. The effect itself was of course well known prior to 1910, with the Lumière brothers having held their first movie screening in Berlin in 1895. However, Wertheimer's train ride marked his epiphanic moment, leading to a new interpretation of the phenomenon and founding the research programme that would become known as *Gestalt psychology*.

Wertheimer continued his investigations by conducting studies on a variety of perceptual phenomena together with Wolfgang Köhler and Kurt Koffka. They observed that stroboscopic motion only occurred when the interval between the disappearance of one image and the appearance of the next was less than 30 milliseconds. They also discovered that when two lights flash alternately and are located close enough to each other, they appear to be a single moving light.[1]

These empirical findings were not, however, the primary reason their work became so influential. Rather, it was their interpretation of such findings that marked a radical departure from mainstream thought. According to standard models of perception, our ordinary experiences are explained by the sensory data we receive and by inferences drawn from past experiences. Yet, as the Gestalt theorists noted, this view fails to provide satisfactory explanations for a number of perceptual phenomena.[2] Why, for instance, do we perceive motion in flickering images when there is none? Or why do two alternately flashing light sources, when positioned closely enough, appear to be a single light, even though they remain visually

[1] A detailed history of Wertheimer and Gestalt theory is offered in Michael Wertheimer (2012, ch. 12) and King and Wertheimer (2005).

[2] A detailed analysis of the deficiencies in the standard approach and the Gestalt alternative is found in Wertheimer (1912).

distinct? As an alternative to the standard view, Gestalt theorists insisted that 'it is not at all true that corresponding to a definite sensation there is a definite stimulus' (Wertheimer, 1944, p. 89). From a range of experiments, they concluded that the experience of an entire visual scene can differ qualitatively from the experience of its individual components. As Koffka famously put it, 'the whole is something else than the sum of its parts' (Koffka 1935, p. 176). The whole, being that emergent unity of perception, is what they referred to as the *Gestalt*, the thing that is more than all parts of a sensation taken separately.

So, if our experiences are built from more than just sensory input, what other factors are in play? For the Gestalt theorist, our cognitive structures impose themselves onto the experiences we have, forming the Gestalt. According to this view, the brain contains certain structures that are prior to sensory input. While sensory data influences our sensations, we too are involved in shaping these sensations by how our brain's processes are structured. This conviction is summarised in Köhler's *Principle of Isomorphism*:

> [The Principle of Isomorphism is] the thesis that our experiences and the [neural] processes which underlie these experiences have the same structure. (Köhler, 1947, p. 344)

While the standard approach holds on to the realist conviction that there exists an isomorphic mapping between our experiences and the external world, Gestalt theory, in Kantian spirit, traces the isomorphism to a relation between experiences and neuropsychological processes. Thus, the Gestaltist's Principle of Isomorphism is precisely the opposite of the realist's: the Gestaltist assumes an isomorphism between neural processing and experience, while the realist assumes one between world and experience.

The standard approach seeks to identify neural mechanisms that detect structure in sensory data, based on assumptions about the nature of the world. The Gestalt theorists turn this around and argue that much of what we experience must originate in the brain itself rather than in the senses. They seek to investigate the subject who, in their view, imposes structure onto the world—structure that may have little to do with how the world actually is. To give another example discussed by Wertheimer (1944), imagine hearing a melody in C major. If you are not a music expert and later hear the same melody transposed to C♯ major, you would likely not perceive a difference. This is remarkable, given that every individual note has changed. Yet the melody as a whole appears identical to almost all of us. How can this be? Supporters of the standard approach to cognition would begin by analysing the data to determine what accounts for the similarity in experience. They might argue that the intervals and relations between the notes remain the same, and that this structural similarity explains why the melody is perceived as unchanged. This reflects the realist strategy of seeking mind-independent regularities and assuming that the brain extracts and represents these accurately. However, such explanations fail to account for findings showing that large variations in note relationships often go unnoticed, while small changes at critical points can render the melody unrecognisable. These studies indicate that the essence of melody perception cannot be reduced to the relations between notes alone. Analysing individual data points is not sufficient to explain the phenomenon. Instead, the Gestalt theorists propose that a proper explanation must involve a holistic account of the *entire* melody and its relationship to the brain.

The endeavour of the Gestalt theorists closely resembles the project of this book. This is evident not only from the Kantian themes they endorse or from the isomorphic principle that echoes idealist thought, but also from their concern with a problem that bears a striking resemblance to what we have referred to as the *clash*. According to Wertheimer, the central problem with which Gestalt theory is occupied is the following:

> We come from the full reality of living events to science, of which we seek clarification, deeper penetration into the core of what is happening; and then we often find, it is true, instruction, information and connections, yet at the end we feel poorer than before. (Wertheimer, 1944, p. 81)

What Wertheimer is saying is that there is an additional element in our experience, namely the Gestalt, which remains unaccounted for in mainstream scientific inquiry. The standard scientific method divides problems into smaller subproblems and treats systems in isolation, thereby failing to address the full range of experiential phenomena. What Wertheimer finds lacking, not only in the brain sciences but in science more broadly, is an account of the cognitive contribution that transforms raw data into Gestalt form. The clash he identifies, therefore, arises from the neglect of neural processing in the interpretation of experience.

10.3 Finding the Rules that Structure Our Perception

When Wertheimer explains that we must treat the sensory input as a whole, he does not mean that the sensory input is not a sum of individual stimuli or that it cannot be treated as separate individual units within the physical domain. Instead, it means to him that once we receive some input information, the entire input is transformed into something new that should be investigated as a whole. The relationship between the individual sensory stimulations and the experience as a whole is what lies at the heart of his cognitive analysis. To understand how these perceived bundles come about, he focuses on visual perception and asks the general question of how our brain creates the experience of *discrete*, coherently arranged objects on the basis of a chaotic accumulation of coloured sensations resulting from the activities of the retinal receptors.

Wertheimer (1923) believes that the brain must follow some rules or laws that govern the grouping processes by which parts of the visual field come to be experienced as composed. He believes that there must be certain properties present in the retinal image according to which the brain decides to form a unit or a group in the retinal image. Through empirical studies, Wertheimer identifies several cognitive principles that guide this grouping behaviour, including factors such as similarity and proximity. The problem Wertheimer investigates is similar to van Inwagen's special composition question, except that van Inwagen is concerned with a mind-independent metaphysics, that is, *real* ontology, while Wertheimer asks questions of compositionality from an idealist perspective.

In contrast to the computational approach to cognition, which focuses on discovering feature detection mechanisms in the brain to reveal the causes behind our visual stimulations, Wertheimer does not investigate how the brain detects basic features such as edges and corners. Instead, he asks how unity comes about as a construct of the brain. He simply presupposes that certain basic units, which he calls ‘elements’, are already given and begins his work directly from there, whereby his methodology relies on subjects reporting on their experiences rather than modelling computational processes.

While this methodology suffers from various limitations, the historical significance of Gestalt theory lies in the vision it offered regarding how the brain might operate and what kinds of questions we ought to ask. One main problem for the Gestaltists was their inability to provide a computational model of how neural circuits could form a Gestalt object or explain how the subjective reports from their perception experiments came about (Jäkel et al., 2016). What they lacked was a model operating in a top-down manner, where higher cortical regions actively impose information on lower ones. This idea was not available to them, as they were unaware of the two-directional computational flow of information in the brain and did not possess the mathematical tools to describe such processes. For these reasons, the realist and computational approach quickly replaced Gestaltist thought, as it offered possible neural encodings that aligned more closely with measurable brain activity.

10.4 Grouping Principles Under the Standard Computational Approach

In their influential work, Palmer and Rock (1994) pick up on Wertheimer’s question of finding grouping principles and integrate it into the computational approach, albeit at the cost of disregarding some of the Gestaltists’ key insights. While they accept the computational detection of features, they also recognise the need for a higher-order mechanism that bundles these features into single units. For this reason, they formulated a model that relied on the standard computational approach of detecting features in the visual field and then providing a computational explanation of how these features are connected into single units, analogous to the rules found by the Gestaltists.

In this approach, Palmer and Rock rightly point out that Wertheimer falsely presupposes the existence of ‘elements’ in the visual field when formulating grouping principles: ‘there are no independent “elements” or “units” to be grouped; there is simply an unstructured image’ (Palmer & Rock, 1994, p. 39). Assuming there to be something like ‘elements’ in the data is very different from the types of objects deployed by computational approaches. Edges and other features that are

Fig. 10.2 Image of a wolf and a moon for which the uniform connectedness principle would work fine

detected by neural circuits are not 'elements' in the sense of being tiny visible colour patches, as the Gestaltists seem to assume, but are rather a stream of numbers that do not necessitate any inherent structure. Based on this, Palmer and Rock formulate a grouping principle called the *uniform connectedness principle*, which does not rely on 'elements' in the visual field but on features computed by neural circuits, from which areas in the visual field can be grouped together. According to this principle, a connected region of uniform visual properties is grouped as one perceptual unit. When we look at Fig. 10.2, for instance, the uniform connectedness principle suggests that we detect the wolf because the surface of the wolf is homogeneous and differs from the background and the moon, which are also both homogeneous in their colour.

A computational realisation of the uniform connectedness principle is feasibly realised by the models of Hubel and Wiesel (1968), using neural circuits that can detect shifts in colour intensities and thereby edges in an image. According to this, the brain would first locate the boundary of an object to then distinguish the bounded domain from the background. This should then result in the individuation of the object that can later be categorised on the basis of previous experiences. What distinguishes the work of Palmer and Rock from the Gestaltists is that ordinary high-level experiences, as well as results from psychological experimentation, can be connected with hypothesised low-level neural mechanisms for processing features in the data.

What we notice is that Palmer and Rock continue the standard cognitive science tradition by assuming that there is a primary, one-directional flow of information processing: light is reflected from the object after which it hits the retina causing a neural response that processes the features in the image from which *after that* the grouping mechanisms occur. They present the grouping mechanisms as a part of the one-directional causal chain of events leading up to the experiences we have.

Palmer and Rock's model is a logical continuation of the standard computational approach, and their all-encompassing model is still, in many circles, seen as state-of-the-art, as I had to witness from review processes of various research articles. But how successful is their model really?

While Palmer and Rock's model does not account for Gestalt phenomena, it also does not explain many or even most of our perceptual experiences, including those of which we are not even aware. Palmer and Rock admit that the uniform connectedness principle has difficulties in explaining how disconnected regions in the visual field can belong to one and the same object if the object is partially covered by another object, visually dividing it. Likewise, the opposite problem exists for objects that are not uniform and visually divide themselves. According to their model, we should experience zebras as a multitude of objects, with each black or white stripe corresponding to a different object. The principle also has difficulties explaining how similar surfaces can still be distinguished, as in the case of animals with camouflage. Imagine we were to apply their principle to the image of the Dalmatian dog (Fig. 9.5 on p. 104). All connected regions that are white would simply be classified as one and the same object. The same holds for all the connected regions that are black. There would be absolutely no way to individuate the dog in the image. We will see in the upcoming chapter how the assumption of a two-directional causal process resolves these problems and how this supports an idealist understanding of metaphysics.[3]

10.5 Rules Based on Information Optimality

A different approach to detecting domains in the visual field that compose into units comes from information theory. The uniform connectedness principle assumes that the different parts of an object's surface are similar. From an information-theoretic perspective, this means that the information we receive from these different parts can be compressed, since their similarity implies a redundancy in description. Many have shown that various aspects of perceptual organisation and structure can be formalised in terms of information redundancy (Garner, 1974) or description length (Buffart et al., 1981; Leeuwenberg, 1968; Leeuwenberg & Van der Helm, 2013; Restle, 1979). In the context of problems concerning our visual ontology, Steve Petersen (2019) nicely illustrates how exploiting information redundancy can result in a useful ontology that avoids many of the puzzles concerning ordinary objects. To describe objects, he relies on Dennett's notion of real patterns that we encountered earlier (see Chap. 6). I remind the reader that 'real patterns' are relatively stable and enduring structural patterns in the data that are non-random and thus allow for more efficient descriptions than random noise.

[3] Palmer and Rock mention the idea that certain developments of top-down neural processing might provide answers as to how Gestalt experiences come about. However, they also state that bottom-up processes are given more attention: 'it seems likely that even higher level factors come into play, such as prior experience of known classes of meaningful objects. Although early Gestalt theorists in fact mentioned this possibility—indeed, past experience was one of Wertheimer's original "laws" of grouping—it has been neglected because of the emphasis on what are now called "bottom-up" processes' (Palmer & Rock, 1994, p. 52).

A string of data like '10111010010' does not seem compressible because there is no clear structure in the string that would allow the data to be described using fewer symbols. On the other hand, '100100100100100100' can be compressed into the shorter expression '6x"100"'. However, this is only possible if the language of description also contains the symbols '6' and 'x' such that the compression of the longer string into the shorter string is possible. If these symbols were not part of the descriptive language, then the data '100100100100100100' would not be compressible either. Likewise, if the language has some symbol, say ♣, that has the identical meaning as the string '10111010010', then the data '10111010010' is in fact compressible by the simple expression '♣'. We see that whether data is compressible or not is always dependent on the language at our disposal, or more generally, the language of the referencing computer or universal Turing machine (UTM). Thus, if a shorter description of some given data can be found in the language of a referencing UTM, then the data is said to be compressible.

Real patterns are representable within a bitmap or pixelated two-dimensional image. From there, the general thought is that if two real patterns jointly constitute a new real pattern that allows for additional compression of the region in which they reside, then it follows that the two get grouped together.[4]

Formally, the information grouping principle (IGP) states that:

IGP Regions in the bitmap are grouped together whenever this leads to the best compression of the totality in which the region resides.

Let us take the case of a ball on a table. Such a situation can be illustrated and visualised by a bitmap of random noise that contains the structure of the objects, being pixelated down to some informational limit, whereby each pixel takes a value from a finite set. Both the ball and the table are thus presented as regions in the bitmap in Fig. 10.3a. Each region can be encoded in terms of the UTM's language of description, just like a subroutine in a computer program. For instance, the pixels in the bitmap encoding the ball can be more efficiently represented when the ball is encoded as a whole rather than being encoded through each pixel individually. According to IGP, the two objects do not group into a single unit, as the grouping does not result in any gain of compression; treating these objects as two subroutines within a routine rather than two disjoint subroutines merely creates additional overhead and lengthens the program's code. However, two parts of a cloud compose and do not exist independently, as their grouping allows for a higher compressibility than their independent existence (Fig. 10.3b).

The provided grouping principle can make sense of certain problems about compositionality within the manifest realm. Petersen discusses different properties of the compositions that are said to exist according to IGP: *maximality*, *minimality*, and *parthood*. The property of maximality, as discussed by Sider (2001), can be attributed to a property F, such as *being a cloud*. F is maximal if and only if large

[4] Petersen believes that he is solving metaphysical problems from a pragmatic perspective. It is not fully clear what his metametaphysical convictions are, but in any case, in the present context, we treat his results as being results about grouping principles at a certain level of reality.

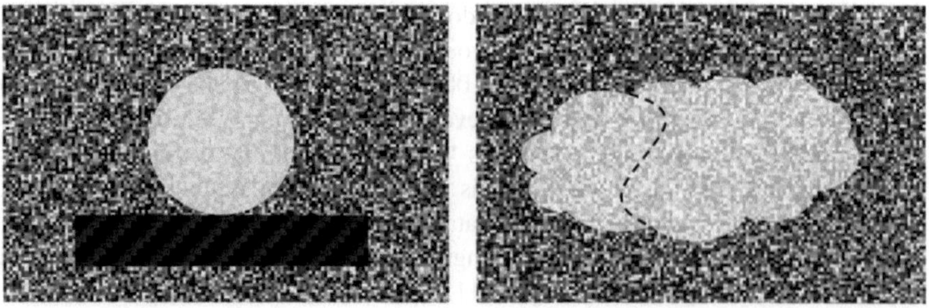

(a) (b)

Fig. 10.3 Examples of two non-composing patterns in the bitmap (**a**), and two regions that do compose (**b**)

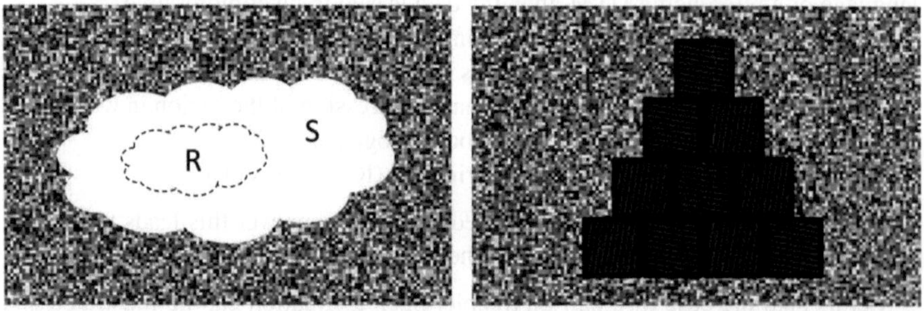

(a) (b)

Fig. 10.4 Examples of IGP-Max (**a**) and IGP-Parts (**b**)

proper parts of an F are not themselves Fs; for instance, a part of a cloud is not itself a cloud. IGP implies the maximality principle, by which subregions do not themselves make up composites:

IGP-Max For two regions *S* and *R* in the bitmap, where *S* fully contains *R*, a pattern over *R* does not compose anything by itself if it can be extended to *S*. Thus, if the conditional information-theoretic complexity for both cases is the same, they are one single composing object.

This principle is a reply to Unger's (1980) problem of the many (Fig. 10.4a), which describes the issue that a multiplicity of possible subsets of an object's basic composing elements, such as the droplets of a cloud, apparently constitute a new object and thereby lead to a population explosion. If IGP is correct, then individual pixels in the bitmap compose if they jointly form a compressible region, and whether or not they compose is not based on our decision, as in Unger's puzzle. The region of composing parts in the bitmap maximally extends to all connected compressible parts and does not leave other connected parts untouched, thereby avoiding a cluttered ontology.

On the other side, minimality is the property that rules out arbitrary compositions, avoiding the conclusion that the ball and the table (Fig. 10.3a) form a new composite object. IGP also implies minimality because it is not the case that any arbitrary summation of pixels results in a gain in simplicity, that is, a smaller information-theoretic complexity.[5]

Furthermore, IGP implies parthood, which is expressed by the following corollary:

IGP-Parts Parts form a whole when the whole is simpler than each of the parts taken separately, but the parts are by themselves simpler than the whole.

Taking the case of a pyramid of bricks (Fig. 10.4b), IGP-Parts aligns with our ordinary intuitions about parthood, namely that both the pyramid and the bricks exist as composite objects. The entire pyramid's data can be encoded in terms of a routine that repeatedly invokes a subroutine encoding for an individual brick. An individual brick is encoded more simply (requiring a shorter program) than the pyramid, but the pyramid's encoding is simpler than invoking the subroutines of all the bricks separately.

10.6 Limitations of the Information Theoretic Approach

The merits of the information-theoretic account of grouping principles lie in its partial alignment with common-sense intuitions. Even though the toy-example cases impressively show how IGP manages to retain some of our intuitions, the theory fails to capture most of our real-world experiences of compositionality. On the one hand, IGP is too restrictive to account for realistic, non-homogeneous objects composed of different textures and colours. A black roof and white walls do not contribute to the composition of a house, as they do not jointly compress the region where they reside. On the other hand, the criterion of composition is too lenient, as it conjoins regions in the data that might allow for compression simply by coincidence. Think again of the Dalmatian dog example, which shares certain similarities with its background such that all connected white regions would have to compose one thing, according to IGP. Furthermore, Petersen himself acknowledges certain difficult cases where IGP seems too permissive, such as the case of two touching coins, which, according to IGP-Parts, compose.

Petersen insists that IGP applies only to *synchronic* composition, which, unlike diachronic composition, does not depend on having a history. Therefore, functional compositions, such as human artefacts, are to be excluded from his analysis. Even granting this, IGP remains far removed from explaining most instances of

[5] The complexity measure referred to here is actually the so-called Kolmogorov complexity. See Kolmogorov (1963).

composition in ordinary experience. Within the information-theoretic approach, rocks and snow do not compose mountains, and land and water do not compose the planet Earth.

10.7 Realism Seems Not to Be the Way Forward

Both the uniform connectedness principle and the information-theoretic approach assume that the brain's primary purpose in the cognitive process is to detect features of the world that can be represented in a more compressed format. This compression is organised by grouping units into data packages. Neither view accounts for the possibility that there might be crucial structural differences between what the world is really like and how it is perceived, since data compression is ultimately about discarding irrelevant information rather than altering its structure.

This realist commitment is quite explicitly stated by both sides. Palmer and Rock assert that the relevance of their proposed principle is straightforward:

> if one presumes that the primary function of perceptual organization is to designate areas in the image that correspond to unified, distinct objects in the 3-D environment . . .. Certainly one of the most important features of what people tend to perceive as a single object is that it is a connected piece of matter, one that could be moved (at least in principle) as a unit. For our present purposes, the important empirical fact about such connected objects . . . is that they tend to have relatively uniform surface characteristics. (Palmer & Rock, 1994, pp. 30–31)

According to them,

> it actually makes good a priori sense for connectedness to be the foundation of perceptual organization, because what people consider to be distinct physical objects are actually *defined* by physical connectedness, at least at a macroscopic level. What really 'goes together' in the strongest physical sense is a single connected piece of matter, not separate ones that happen to be near each other in space. (Palmer & Rock, 1994, p. 32)

These quotes exemplify how Palmer and Rock are ultimately committed to a realist view, assuming that our experiences of object structure directly reflect facts about the world. Although the uniform connectedness principle is innately encoded in our brains, it is taken to mirror the actual structure of the external world.

Petersen even explicitly mentions that he seeks an 'objective' and 'realist' (p. 1121) answer to the problem of compositionality as far as possible. The fact that IGP is always UTM-relative, since *any* finite data, x, can be compressed relative to *some* UTM (for example, by storing x as a hard-coded value in the language of description), is of little concern to him. Petersen is not, however, precise in explaining what his realism amounts to.

The general concern with both of the discussed approaches is that they fail to recreate most of our ordinary experiences by seeking criteria for how things lie in the world and how the brain processes them accordingly. Their approaches simply do not seem to yield the results we ultimately seek. Both approaches take seriously the problem Wertheimer poses about finding grouping principles, but in contrast to the Gestaltists, the criteria they seek are motivated by being in some way connected to the physical world. While the Gestaltist believes in the Principle of Isomorphism that decouples structure in perception from structure in the world, the two later-developed approaches seek a structural connection between both neural representations and the mind-independent world. We will see how idealist approaches are more successful in producing ontological frameworks that explain our manifest experiences.

Summary
Gestalt psychology was one of the first movements to oppose the standard view in psychology and the cognitive sciences, observing that the brain significantly shapes the data we receive through our senses. Thinking of the brain as the constructor of our structural experiences, the movement is closely connected to idealist thought. It was not successful, however, as it failed to offer precise computational principles and could not identify the neural underpinnings of its theories. While the standard realist approach to grouping principles manages to provide some account of how our visual ontology acquires its structure, it remains very limited in its explanatory power.

Chapter 11
From the Concept of Life to Neural Processing

In previous chapters, we encountered the standard cognitive science approach to revealing the neural code. We saw that, in essence, this approach is committed to hard-wired feature-detecting circuits that classify identified objects based on similarity and can be characterised as a one-directional causal process of events that primarily extracts information from the outside world. Let us now turn to a theory that has completely revolutionised our view of cognition and has challenged each of these assumptions of the standard approach. While there are many possible starting points for explaining this radical new perspective, I choose a narrative that begins not with neuroscience or the brain, but rather with an understanding of life.

11.1 What Is Life?

In an inquiry concerning the nature of living beings, the physicist and Nobel Prize laureate Erwin Schrödinger seeks a distinction between living and non-living things in the world. There seems to be a vital difference between physical objects that are subject to experimentation in physics laboratories and living beings as we know them, including everything from the human species to a unicellular organism investigated under a microscope. Schrödinger claims that these differences amount to their physical order or *entropy*. According to the second law of thermodynamics, all systems tend to increase in entropy over time and thus decrease in order. A sugar cube in water will eventually dissipate, which means it loses its regular crystal structure. A cup of coffee will eventually cool down because heat does not rest at one localised position but flows over time towards regions of lower temperature. A landscape is eroded by water flow and winds. As Schrödinger observes, living beings are, in this respect, different. They manage to preserve their order even when faced with environmental changes.

A. C. Schwaninger, *Cognitive Metaphysics*, Synthese Library 529,
https://doi.org/10.1007/978-3-032-05420-3_11

Understanding life as not falling into decay raises some difficulties. First, a paradox seems to occur. If living beings are composed of matter and matter is driven by the laws of physics, including the second law of thermodynamics, how can they resist decay? Of course, there is a simple answer to this question: one must consider the system as a whole and not the living organism in isolation. The system as a whole is still heading towards decay, even if a subsystem of this whole is not. Living beings have a very efficient metabolism that helps them extract energy from food and external heat to maintain their physical boundaries, at least for a long period of time. By nourishing themselves, they gain energy from ordered structures, allowing for self-preservation at the cost of decreasing order elsewhere.

The second worry is that Schrödinger's answer to the question of what life is seems to beg the question in an almost tautological way. Whatever is alive has, by definition, not decayed; otherwise, it would be dead. Saying that life is whatever does not decay (for a long period of time) is simply restating the contrast between life and death. Furthermore, most systems we conceive of as living survive, at best, a hundred years. This is nothing compared to the lifetime of the Swiss Alps, which have existed for millions of years. These concerns can be avoided if life is understood as an activity, and living systems are defined as those that engage in this activity, namely the activity of counteracting decay. In contrast to rocks and mountains, animals, plants, and other living creatures actively resist decay in the face of changing surroundings. Some forms of activity are simply better at this than others, and any system that is particularly good at countering decay is then conceived of as being alive. While rocks and mountains are clearly not living under this definition, oil droplets in water, which occasionally appear to merge and be torn apart by the water's movement, also do not engage in any activity that helps them sustain themselves.

Living systems actively sustain themselves by their capacity to be in (or *occupy*) a vast number of different bodily *states*, such as eating, hunting, seeking shelter, or holding a cup of tea while riding a bicycle. This ability to exist in many different states is necessary, since otherwise it would be impossible to react appropriately to different environmental changes. For this, the different parts of the system's body must interact and coordinate with each other such that the system as a whole reaches beneficial states. The more states a system can occupy, the more likely it is that it can react appropriately to a new environment. A stone that always occupies only one single state cannot react to or avoid falling from a cliff and breaking apart. The ability to react requires the ability to convert from one bodily state to another. As we have seen earlier (Fig. 9.2 on p. 97), while the most complex system is always the system that is non-compressible, such a system lacks the order needed to sustain itself. Thus, if an organism has only a limited capacity to occupy different states, it might be unable to react appropriately to certain environmental changes. However, if it occupies too many states, it can become too complex and thereby fail to organise itself appropriately, since some of its parts behave in an uncoordinated manner. While some degree of order might still persist for longer periods of time in inanimate substances, such as the planets orbiting the sun or a mummy preserved in a tomb, the rate of changes in the systems' states is typically very low and order cannot be maintained once surrounding conditions change. While a rock that falls into a stream of water becomes helplessly subject to erosion, a land animal might be able to help itself by bringing itself into the state of 'swimming' and thereby getting back to land.

Up to half a century after Schrödinger's death, the theoretical neuroscientist Karl Friston developed a theory that takes these inquiries into what it means to be a living system as the theoretical starting point for his own general theory about the brain.[1] Following Schrödinger's views, Friston's theory defines a living being as a system that actively maintains order for the purpose of survival, or rather *homeostasis*, the organism's equilibrial state in which conditions of optimal functioning are in place for preserving its boundaries. Using this definition as a starting point, Friston concludes that a living being must not merely *contain* mechanisms that allow it to persist within the bounds of an equilibrial state, for example, by regulating its temperature. Rather, the organism as a whole *is* a set of states that manage to maintain themselves within certain bounds either by adaptation in the form of transitioning from one state to the next or by changing the environment. For example, when infected by a virus, the transition into a state of temperature increase (a fever) can help survival, but when standing in a fire, the agent must occupy states that alter the environment itself, such as dumping water over it.

One useful abstraction that allows us to conceptualise these ideas is to describe the states of a system as regions in *phase space* (or *state space* when dynamics is absent). Phase space is a term widely used in disciplines concerned with dynamic system behaviour and refers to the mathematical space in which all possible states of a system can be represented. In Fig. 11.1, a frictionless pendulum is shown that is in a constant state of motion, which can easily be demonstrated to trace an ellipse in a phase space spanned by the momentum (p) and position (x) coordinate axes. Real-world systems are usually much more complex due to more intricate motions and the fact that the dimension of phase space is twice the number of degrees of

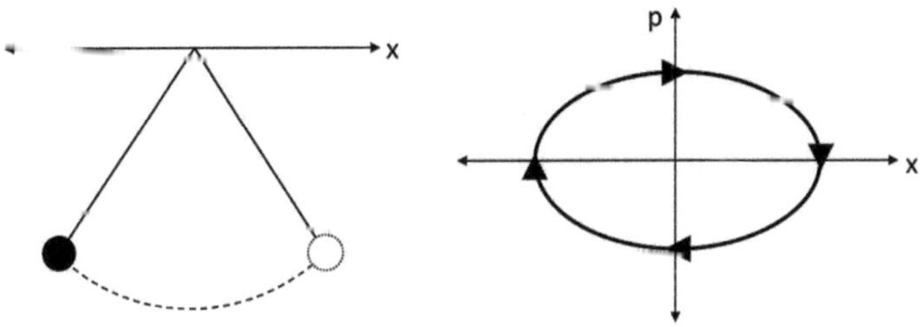

Fig. 11.1 A harmonic oscillator such as a frictionless pendulum (left) has an elliptic trajectory in phase space (right)

[1] This does not mean that the way he developed the theory used the question about life as a starting point. However, it is possible to derive all critical parts of his theory from a notion of life. See Friston (2012a) on the historical details of how he developed his theory.

freedom. A system of 100 particles moving independently in 3-dimensional space is thus represented in a 600-dimensional phase space.

In the pendulum example, the system's possible states are limited to a very small region in phase space. Only by cutting the thread would regions with x greater than the pendulum's length become accessible. Likewise, in order to avoid death, a living system is limited to a relatively small region in phase space consisting of different states that are frequently occupied. A system that occupies the same states over and over again is said to be 'ergodic', which is a good approximation of living beings (Friston, 2013a, §4). Accordingly, a 'living agent simply *is* a set of states that maintain themselves within certain bounds, the bounds that describe the conditions necessary for their own survival' (Clark, 2017, p. 3). All parts of a living system actively manage themselves and each other to occupy the states in phase space that are beneficial for survival, and they try to ensure that those states remain bounded within a small region of phase space.

11.2 Boundaries of Biological Individuals and the World Beyond

Where do the physical boundaries of a living organism lie? One might think that the individual organism is simply the connected matter forming the system that preserves itself. One might point at a dog and suggest that its fur, its teeth, its skin and all the connected parts that can be seen from outside constitute the outer boundary of this living individual. However, the question cannot be answered this easily. Consider, for instance, that the dog itself consists to a large extent of bacteria, many of them living in its digestive tract, on which the dog's survival depends. Are these microorganisms now a constitutive part of the dog, or do they exist as independent living beings, even though their own survival depends again on the environment provided by the dog?

Both the boundaries around a dog and the boundaries around a microorganism in the dog are boundaries of living organisms. A useful mathematical way of defining a living system's boundary is given by the mathematical formalism of *Markov blankets* (Kirchhoff et al., 2018). The Markov blanket draws a partition between the system's internal (e.g., state of the heart pumping) and external states (e.g., environmental states), where the blanket itself consists of the states that separate the other two kinds of states. The blanket is defined to be the boundary separating the causal interaction between the inner world of the organism and the outer world. This separation is useful because it makes a distinction between those states that the system under consideration can control in order to sustain itself and those that lie either completely beyond the system's reach or are at least not always under its control. When the environment is cold, the body cannot do much about the general climate, but what it can do is activate certain muscle movements, creating a shiver that keeps it warm. It can also influence the environment by changing it, for example, turning up the heater, but it has no constant control over the heater. The notion of a Markov blanket was originally proposed for the description of Bayesian networks and defines the conditional independencies between two sets of states within a network (Pearl, 1988).

The Markov blanket formalism is useful as it describes any boundary of any self-organising system at any possible scale. For instance, a cell has, on the one hand, intracellular states such as the protein formation activity. These internal states can affect the states on the Markov blanket, in this case the cell membrane, which contains sensory receptors to retrieve information from the environment and also acts upon the environment in ways that allow the cell to alter its position. Essentially, any ergodic system can be characterised in terms of a Markov blanket to distinguish inside from outside. While a dog as a whole has a Markov blanket, so does its brain, and so does each cell in the brain. One therefore sometimes talks about *nested* Markov blankets to clarify that one is speaking about Markov blankets within Markov blankets.

The universal applicability of Markov blankets should, however, not be exaggerated. It would be false to think that 'we should be able to describe the universe in terms of Markov blankets of Markov blankets, and Markov blankets all the way up, and all the way down' (Ramstead et al., 2018, p. 5), because the universe is probably not an ergodic system. An ergodic system, by definition, revisits its states from time to time, which is not generally true for the universe. The universe seems to have only one Big Bang, from which it is always expanding. It does not seem to revisit states it has occupied earlier. In fact, nothing is truly an ergodic system. Ergodicity is merely an approximation, in the same way that it is only an approximation that living beings defy the second law of thermodynamics. They might do so for a certain period of time, but eventually all things will end in the heat death of the cosmos.[2]

The mathematical description of the Markov blanket characterises the external states of the world as conditionally independent from internal states, meaning that the internal states only have indirect access to the external states through the 'veil' of the Markov blanket (Clark, 2017; Friston, 2013a). This is a mathematical reflection of the underdetermination problem, which tells us that the causes of our sensations must be inferred and cannot be accessed directly. As Jakob Hohwy (2016) says, there is an *evidential boundary*, separating the brain from everything else. This boundary opens the door to scepticism and challenges naïve realism, the view that our perceptions present us with a mind-independent reality.[3]

The biological systems that we are interested in can be summarised as having Markov blankets and as actively preserving their existence under environmental changes. What, then, does this activity amount to? The activity in which a living system engages is that of trying to determine its best response, in terms of its actions

[2] Notice that some physicists believe that there are other destinies for the universe. For instance, according to Smolin (2014), gravitation does not allow the universe to reach thermal equilibrium. This does not, however, affect the overall account presented here.

[3] For a detailed definition and discussion of naïve realism, see Beck (2019).

and adjustments of internal states, in order to sustain itself, given that information about the world is limited. From the limited information it receives through the senses, it must infer the appropriate decisions that will enable its survival.

11.3 Bayesian Probability Theory

In the old, standard cognitive science approaches, the problem of underdetermination is solved by hard-wired mechanisms that detect specific structures such as edges or motion direction. In these models, a specific input yields a specific output, and apart from biological noise, the circuits respond in a determinate, all-or-nothing fashion.

The modern understanding of how the brain addresses the problem of underdetermination focuses on modelling perception through a *Bayesian probability* perspective. This approach treats perception as a matter of making probabilistic inferences based on prior beliefs. In a realist interpretation, these are beliefs about the world that motivate certain actions. In contrast, an idealist interpretation treats them as beliefs about the usefulness of an action, with less concern for the world as it might exist independently of the mind. Since the realist interpretation is easier to grasp, I will explain the Bayesian view from that angle. However, the idealist version involves essentially the same structure, with the only difference being the absence of claims about mind-independent objects; instead, it is concerned solely with a constantly changing world as it appears to the mind.

The central idea of the Bayesian perspective is that all our perceptual beliefs are ultimately probabilistic. The best way to model the formation and revision of these beliefs is to use mathematical tools from probability theory, which allow us to describe how beliefs evolve as new information is acquired.

Placing probability theory at the heart of neural processing is motivated by the mathematical fact that any system forming beliefs based on reason and rationality must be structurally equivalent to probability theory. A system whose credences violate the laws of probability will end up holding irrational beliefs.[4] The connection between mental thought and probabilistic inference was already recognised by the mathematician Pierre-Simon Laplace:

> One may even say ... that almost all our knowledge is only probable; and in the small number of things that we are able to know with certainty, ... the principle means of arriving at the truth – induction and analogy – are based on probabilities. (Laplace, 1812, p. 1)

For Laplace, the entirety of human knowledge is tied together with probability theory.

But why probability rather than logic? Is logic not the tool of all reasoning? Consider a proposition P = 'if A is true, then B is true'. If A is true, then classical

[4] This is shown by Cox's theorem, an overview of which is provided by Horn (2003), and is illustrated by the well-known Dutch book argument.

logic allows us to deduce from P that B must also be true. But what can we infer from P if we are told only that B is true? According to classical logic, we cannot infer anything, since P tells us nothing about A given that B is the case.

Suppose P = 'if John has a dog, then he has a pet'. If we are told only that John has a pet, classical logic comes to a halt: it cannot tell us anything about whether John has a dog, a cat, or some other animal. Probabilistic reasoning is different. If we know nothing about John, we have little reason to believe that he owns a dog. However, if we are told that he has a pet, then the probability of him having a dog increases. The new information makes the dog hypothesis more plausible. This illustrates how probability theory extends classical logic by allowing inference under uncertainty and by covering a wider range of inferential contexts.[5]

This new perspective on how to model neural processes is accompanied by a new perspective on what probabilities are and what we mean when we speak of probable and improbable events. Or rather, the perspective arises from rediscovering an old view on a mathematical concept and applying it to modern scientific contexts. The key insight is that probabilistic behaviour must not be interpreted as random events occurring *in the world* but should instead be understood as a reflection of our own epistemic limitations. Most statisticians in the twentieth century follow the *frequentist* tradition, according to which probabilities describe the frequency of events in repeated, randomised trials. Saying that the probability of a fair coin landing heads is 0.5 (or 50%), for example, means that if we were to repeatedly toss the coin and count the number of times it lands heads, the average ratio of heads to total tosses would be 1 to 2 (=0.5). This interpretation of probabilities was already foreshadowed by Aristotle, who said that 'the probable is that which for the most part happens'.[6] However, it was never accepted by all scholars. Jacob Bernoulli, for instance, who worked on probability theory around 1700, noted that interpreting probabilities in this way depends on knowing the number of all possible events. In most practical situations, we do not know this number. For example, we have no clear idea how many distinct diseases there are, which prevents us from calculating the probability of falling ill.[7] This is not the only problem with frequentism. If probabilities express frequencies in outcomes of repeatable random events, what does it mean to say that 'the probability of a nuclear power plant exploding is very low' or that 'it is very likely that the universe started with the Big Bang'? The first statement seems epistemically inaccessible, since we cannot conduct repeated trials and have only a small historical sample. The second seems even more problematic, because the Big Bang is not a repeatable event—it either occurred or it did not. Its truth cannot be tested through repeated experiment.

The *Bayesian* view on probabilities avoids these problems by interpreting the concept as referring to incomplete human information. The only reason why we see a coin fall on heads or tails at random is because we have no access to the

[5] A detailed discussion of probability theory as an extension of logic is given by Jaynes (2003).

[6] See *Rhetoric*, Bk 1, ch. 2; discussed in Franklin (2001, p. 110).

[7] See Jaynes (1986) for a brief historical overview and arguments in favour of Bayesianism.

dynamic information involved at the moment the coin is tossed. The absence of this information makes the outcome *appear* to be indeterministic, despite the fact that every coin flip has a deterministic outcome if we were able to precisely compute its motion based on detailed knowledge of the initial throwing conditions and air resistance. It would be a fallacy to believe that the world is *ontologically*—that is, in a mind-independent sense—statistical merely because we observe indeterminate behaviour.[8]

Thomas Bayes, an eighteenth-century parson, envisioned such an interpretation of probabilities, but it was Laplace who discovered a useful mathematical formalism that he applied to various scientific domains like astronomy and meteorology. For two events (or propositions) A and B, Laplace showed what is today known as *Bayes' Theorem*:

$$P(A|B) = \frac{P(B|A)\,P(A)}{P(B)}$$

$P(A|B)$ is the *conditional probability* and expresses the likelihood of an event A occurring, given that B has occurred or is the case. The theorem states the relationship between the probability of an event conditioned on a second event and the probability of the second event conditioned on the first. I like to refer to this as the 'common sense theorem' because it follows trivially from $P(A \cap B) = P(A|B)\,P(B) \stackrel{!}{=} P(B \cap A)$, and because it illustrates much of how our common-sense reasoning works. Although there are many cases in which people fail to reason rationally, especially when emotionally biased, there are probably many more cases in daily life in which our reasoning aligns with Bayes' rule.[9]

To understand Bayes' Theorem better, we must consider it in action. Imagine you are walking down the street and suddenly see a shadow that appears to be coming either from a female lion or from a house cat standing behind you. Is it more likely, and thus more reasonable, to think that there is a lion or a cat behind you? If we assume that the shadow resembles both a lion and a cat equally well, then the probability of seeing this shadow, given that it comes from a lion, is equal to the probability of it coming from a cat:

$$P(\text{Shadow}|\text{Lion}) = P(\text{Shadow}|\text{Cat}) = 0.5.$$

[8] This is what Jaynes calls the 'mind projection fallacy' (Jaynes, 1989).

[9] There is an enormous literature on systematic irrational behaviour in the social sciences. For a starter, one might turn to Kahneman (2011). There is likewise a great deal of evidence that people reason on the basis of probabilistic inference and prior knowledge (Tenenbaum et al., 2006). I believe the latter is more common, since all of our perception relies on Bayesian reasoning, as we will see in the following.

The probability of seeing such a shadow is irrelevant to our analysis, and we can treat it as a constant:

$$P\,(\text{Shadow}) = \frac{1}{C_1} > 0.$$

The *prior* probability of there being a lion behind you might be 0.0001 in South Africa and 10^{-14} in Europe. Assuming you are located in South Africa, the *posterior* probability that the shadow belongs to a lion is

$$\begin{aligned} P\,(\text{Lion}|\text{Shadow}) &= \frac{P\,(\text{Shadow}|\text{Lion})\,P\,(\text{Lion})}{P\,(\text{Shadow})} \\ &= C_1 \cdot 0.5 \cdot 0.0001 \\ &= C_1 \cdot 0.00005. \end{aligned}$$

The posterior probability that the shadow belongs to a cat is

$$\begin{aligned} P\,(\text{Cat}|\text{Shadow}) &= \frac{P\,(\text{Shadow}|\text{Cat})\,P\,(\text{Cat})}{P(\text{Shadow})} \\ &= C_1 \cdot 0.5 \cdot 0.001 \\ &= C_1 \cdot 0.0005, \end{aligned}$$

if we assume that it is ten times more likely for there to be a cat walking along the street than a lion, making the prior $P(\text{Cat}) = 0.001$. Since $P(\text{Lion}|\text{Shadow}) < P(\text{Cat}|\text{Shadow})$, it is, by Bayesian reasoning, rational to believe that there is a cat, not a lion, behind you, given the visual evidence of the shadow.

What makes Bayesian reasoning so useful is its ability to update its model given *new* information. Given the information of the presence of the shadow, we have computed that it is reasonable to feel safe. But how does this change when you hear a roar from behind? In this case, we apply Bayes' formula again, this time inserting $P(\text{Lion}|\text{Shadow})$ as the prior probability, since the term computed above represents the state of our knowledge before receiving the new information—the roar. We can assume, for instance, that the probability of a lion roaring is very high and that hearing a cat roar is very unlikely: $P(\text{Roar}|\text{Lion}) = 0.9$, $P(\text{Roar}|\text{Cat}) = 0.0001$. As before, $P(\text{Roar})$ can be treated as a constant $\frac{1}{C_2} > 0$. We then compute the probabilities as follows:

$$\begin{aligned} P\,(\text{Lion}|\text{Roar, Shadow}) &= \frac{P(\text{Roar}|\text{Lion})P\,(\text{Lion}|\text{Shadow})}{P(\text{Roar})} \\ &= C_2 \cdot 0.9 \cdot C_1 \cdot 0.00005 \\ &= C_1 \cdot C_2 \cdot 0.000045 \end{aligned}$$

$$
\begin{aligned}
P\,(\text{Cat}|\text{Roar, Shadow}) &= \frac{P(\text{Roar}|\text{Cat})\,P\,(\text{Cat}|\text{Shadow})}{P(\text{Roar})} \\
&= C_2 \cdot 0.0001 \cdot C_1 \cdot 0.0005 \\
&= C_1 \cdot C_2 \cdot 0.00000005
\end{aligned}
$$

From these calculations, we see how additional information provides rational grounds for updating our beliefs. After including the new evidence, the probability that a lion is standing behind you is now significantly higher than that of a cat. Your brain should therefore activate an appropriate response: run![10]

The Bayesian principle for reasoning dictates that we should assign our credences $q(A)$ about a proposition A to the posterior $p(A|B)$ when the information B is given. Our senses are constantly receiving new information, and much of the brain's task is to relate this new information to prior information. The ease with which the Bayesian framework allows us to conceptualise adjustments of the brain's internal models in light of new data is the primary reason why the Bayesian approach has become central to modelling and understanding the brain.

11.4 Sustaining by Predicting and Minimising Free Energy

Having now established a formal understanding of how the brain reasons with data, we can ask: what exactly do living systems do in order to sustain themselves? Friston approaches the question of what it means to be a living being—or to be engaged in the activity of living—from a purely mathematical standpoint. What he finds is that if a system acts in order to exist, it must act to avoid states of low probability. This means it must avoid entering states that are highly *surprising*.

But what does this mean in concrete terms? Any action a system performs, and any update it makes to its internal states, is ultimately aimed at moving toward regions in phase space that it has ideally already visited and intends to revisit repeatedly. For this to happen, the system must not be surprised by the environment. If the environment changes in a way that the system is unprepared for and does not know how to respond to, it experiences surprise. In the worst case, this can lead to death, as the system fails to cope with the unexpected situation. Of course, this does not mean that every unsurprising situation enables the system to maintain itself. Even if you are not surprised about what will happen after falling from an aeroplane without a parachute and are fully aware of your acceleration towards the earth's surface, there is little you can do to prevent entering a region of high probability in phase space—one where your Markov blanket is inevitably torn apart.

[10] It is assumed that the constants are greater than 1, which follows from the fact that $\frac{1}{C_i} \leq 1$, since probabilities are by definition less than or equal to 1.

Even if not always successful, living systems must at least attempt to avoid surprising events on average in order to sustain themselves. Friston showed that for a system to achieve this, it must make *predictions* about what the world is like and how it will be affected by the world, so that it can adapt and behave accordingly. This is a rather unintuitive finding if we compare it to how we usually think about the body. According to common sense, one would assume that a living system must primarily be able to *react* to changes in its environment. However, although it remains true that a system must know how to react, this is not its primary mode of functioning. Instead, it mainly aims to predict future states of affairs and behave in accordance with these predictions. While this is still a kind of 'reaction', it is a response to the system's own predictions rather than a passive reaction to an external event.

Prediction comes in various forms, illustrated by a range of examples. When we follow our New Year's resolution and start going to the gym more regularly, our muscle mass increases—not because this is an immediate reaction to the current situation, as an increase in muscle volume tomorrow will not retrospectively change our performance in lifting weights today. Rather, our body makes a prediction that it will need greater muscle strength in the future, since similar situations requiring such strength are likely to reoccur. Another example is the state of fear: when we feel afraid, our blood starts pumping faster and adrenaline is released in preparation for fleeing. The body shifts into a state that is not immediately needed, but could become necessary moments later. A further case is genetic inheritance through natural selection. Genes determine nearly all aspects of an organism's structure and are, for the most part, immutable. It is therefore essential that genes encode the right features and abilities that will help the organism sustain itself in its environment. The genetic makeup of humans, for instance, is clearly not suited for life two hundred metres below the ocean surface. Genes, in this sense, form a prediction about the capacities an organism will need in a particular environment, and natural selection is the process through which such predictions are refined over time.

According to Friston, any living system is constantly altering its internal states and performing actions in ways that, on average, minimise prediction error. This, in turn, maximises the expected likelihood of remaining within the bounds of existence. One might worry that such a system would simply retreat into a dark, isolated room, where all encounters become predictable. However, this concern is unfounded. The world extends far beyond the confines of that room, and the longer an organism remains inside, the less capable it becomes of predicting what occurs outside. Eventually, it might not even be able to tell whether it is day or night.[11] The key idea is that a living system aims not merely to possess a model of the world but to *be* that model. In doing so, it maintains its own existence. A good model allows the system to respond effectively to external changes by making accurate

[11] The famous dark-room thought experiment was introduced in the early stages of developing predictive processing and the free energy principle, and was initially used as a counter-argument to the theory. It is now widely accepted that it poses no serious challenge. For detailed discussions, see Baltieri and Buckley (2019) and Friston et al. (2012).

predictions. This is why living beings typically exhibit curiosity and an exploratory drive, in addition to avoiding threats to their survival.

In more formal terms, what the organism is doing when making predictions is minimising its so-called *free energy*, which measures the discrepancy between its predictions (or internal model) and the actual sensory inputs it receives. Free energy can be thought of as a numerical expression of how wrong the organism's expectations are about its environment. If the world behaves as predicted, free energy is low; if something unexpected occurs, free energy increases. This process reduces uncertainty and helps maintain a stable state. By constantly adjusting its beliefs and behaviour to align with what is happening around it, the system avoids being thrown into disarray by surprising events. Friston's (2010) account is referred to as the *free energy principle* (FEP). The principle states that any self-organising system that exists in a non-equilibrium state within its environment must minimise its free energy in order to sustain itself. Or, as Friston explains: the FEP states that 'any self-organising system that is at equilibrium with its environment must minimise its free energy. The principle is essentially a mathematical formulation of how adaptive systems (that is, biological agents, like animals or brains) resist a natural tendency to disorder' (Friston, 2010, p. 127).[12] Minimising free energy amounts to minimising the long-term average of prediction error a system makes about incoming sensory data, given its internal model (Friston, 2010; Friston & Stephan, 2007). The notion of a Markov blanket allows us to describe the boundaries of such a free energy minimising system, such as a living being.

Let us briefly look at the FEP from a more mathematical angle. When we are given some sensory data x, we have seen that it is rational to assign our credences $q(w)$ about some event w that lies behind the Markov blanket in accordance with the Bayesian posterior: $q(w) \leftarrow p(w|x)$. This approach works well when the credences are sufficiently accurate in relation to one another. In the earlier example, we computed that the probability of there being a lion was greater than that of a cat, given the shadow and the roar. We did not, however, compute the precise probabilities but only the terms up to a constant, since the priors of the data $p(x)$ (in that case $P(\text{Roar})$ and $P(\text{Shadow})$) are generally infeasible to determine. For this reason, computing the exact posterior $p(w|x)$ is typically intractable, even though it would be ideal for the organism to assign absolute credences. The solution is to *approximate* $p(w|x)$ with a model $q(w)$ rather than finding an exact solution. If the organism had a perfect internal representation of the world, then all its predictions would also be perfect. In this ideal case, we say that *surprise* is zero. In all other cases, the organism constructs an approximate model to try to predict the future states of the world as accurately as possible.

The ultimate goal is to minimise surprise via making correct predictions. Given some sensory data, surprise is a measure of the amount of information contained in a given event. When an event is expected to occur, the amount of surprise will be

[12] For the mathematical details, see Friston and Kiebel (2009), Friston (2009, 2010, 2013a), Friston et al. (2016), and Friston et al. (2006).

low, whereas very unlikely events result in a high degree of surprise. This notion of surprise also matches our subjective experience of surprise. Consider the following example: when Joe wakes up in the morning and follows his daily routine of taking a shower, brushing his teeth, and so on, the amount of surprise he experiences is very low, resulting in almost no adaptation of his inner mental model of the world. His brain makes little to no model adjustments by changing synaptic weights or connections (i.e. neural plasticity), because for Joe's survival purposes, there is no reason to make any such adaptations. If, for some reason, his wife asks him whether he has locked the door after leaving the house, he may be uncertain. This is because his daily routine has become so mechanical that the information about him locking the door has triggered no surprising event and did not, therefore, result in any significant change in Joe's internal model of the world. However, if Joe had been surprised by the door's lock being broken, he would be able to report on it.

Given its finite resources, an organism's model of the world cannot perfectly correspond to the world. The (variational) free energy is an expression of the inaccuracy of the organism's model approximation. For the mathematically inclined reader, I briefly outline the mathematics behind this quantity. It will not, however, be crucial for the rest of the book.

Mathematically, the free energy involves two components of inaccuracy: one described by a measure of surprise, and one by a measure of how well the model overlaps with the true distribution. For a model q and sensory input s, surprise is mathematically described as $-\log p(s|m)$.[13] This measures the amount of unpredicted information and represents the kind of error the organism tries to avoid on average. The difference between the two distributions p (the true posterior) and q (the organism's internal model) is measured by the *Kullback–Leibler divergence*, $D_{\mathrm{KL}}(q, p)$. This term provides a distance between two distributions and penalises cases of *overfitting*, such as when the model fails to generalise and instead fits too closely to specific data. The free energy can then be summarised as the sum of both terms:

$$F = \underbrace{D_{\mathrm{KL}}(q, p)}_{\text{Kullback–Leibler divergence}} \underbrace{-\log p(s|m)}_{\text{surprise}}.$$

The free energy is known to the system, as it describes the general inaccuracy of the model the system infers based on the sensory data it receives. The system does not, however, know the exact source of this inaccuracy, since both terms—the Kullback–

[13] Surprise S is a logarithmic measure to compute the amount of information of an event and was famously proposed by Shannon (1948), even if he did not refer to it by this name. More formally: Let (Ω, F, P) be a probability space. The surprise of an event $f \in F$ is given by $S(P(f))$, where the function $S : [0, 1] \to [0, \infty)$ is characterised by: (a) $S(1) = 0$, ensuring that a certain event generates no surprise; (b) if $S(x) > S(y)$, then $x < y$, so less likely events result in more surprise; (c) S is continuous, such that small changes in probability do not yield major jumps in surprise; (d) for two independent events a and b: $S(xy) = S(x) + S(y)$, ensuring the additivity of independent events. The logarithm satisfies these conditions such that for some constant c: $S(x) = -c \log x$.

Leibler divergence and the surprise—are themselves unknown. By altering its guess for the best model q, the system can reduce its free energy and, in doing so, also reduce its average surprise. Thus, by making better approximations, the organism can converge on the model distribution that minimises free energy. This distribution then determines its credences. The Bayesian principle is therefore refined by the FEP, which states that the model distribution minimising free energy is the model of choice:[14]

$$q(w) \leftarrow \arg\min_q F.$$

11.5 Predictive Processing

Minimising free energy is an indirect way of minimising surprise. This happens by taking appropriate actions, like looking in the direction in which most novel information is provided to minimise surprise in the future, and by adapting internal states to provide accurate representations for future events. The latter is not possible for all organisms, since not all organisms have the appropriate mechanisms to be adaptive. Thus, not all organisms minimise free energy in the same way. As we have seen, there are many possibilities for making predictions about the data coming from the environment that do not involve direct internal changes. It might just be that the predictions are hard-coded in the DNA and not adaptable to environmental changes—at least not for an individual organism in isolation with one given set of genes.

The crux of the matter, however, is that the FEP formalism allows us to make sense of what the central information processing unit of humans—the brain—is doing when it deals with our senses. The brain has internal brain states that affect the world by adjusting its boundary states to trigger certain actions or reflexes that change the environment and the data it receives from it. The information about these changes is received at the brain's Markov boundary via the senses, which again influences the inner brain states. The brain is more advanced than other systems with regard to the techniques by which it minimises free energy. While simple single-cell organisms might be unable to alter internal representations and can only minimise free energy by wiggling from one direction into another, the brain can adapt its internal representation in a highly complex manner. Under the assumption that the human brain is particularly good at minimising free energy, the FEP offers a theory of how the brain works.[15] Despite its wide applicability, the FEP is likely the most advanced theoretical framework of the brain to date and gives a unified account of

---

[14] See Mann et al. (2022) for a detailed comparison between Bayesian inference and free energy minimisation.

[15] The idea that actions are performed to change the data that are sampled from the environment is called active inference (Braitenberg, 1986; Powers, 1978, 1973).

learning, action, and perception. It is said to lie at the heart of any respectable brain theory (Friston, 2010).

According to Friston, general principles about the neural code can be derived from first principles of what it means to be a living being. Historically, however, the way Friston interpreted and developed the FEP was different. He first suspected that merely some isolated brain regions implement the FEP (Friston, 2003). A few years later, he moved on to extrapolate his claim to a more radical version: that almost everything the brain is doing involves minimising free energy (Friston et al., 2006), and that the FEP provides the first unified theory of the brain. It was only much later that he went a step further in claiming that the FEP applies to *any* biological system (Friston, 2012b).

While the FEP is a guide to what the purpose of neural computations is, namely minimising free energy, it is unspecific about the concrete realisation. This is where *predictive processing* (PP), as a novel framework of how the brain works, comes in. PP models offer a concretisation and specific account of how the FEP is realised in humans.[16] While the FEP is a framework that implies the general principles of PP, PP models offer a concrete understanding of the realisation of these principles. They were developed independently from the FEP and were only later found to be generalisable by it.

To understand PP, let us consider some physiological facts about the brain. Since Santiago Ramón y Cajal's seminal work on the neocortex, which includes drawings of cortical lamination (Ramón y Cajal, 1911), we know that physiologically the neocortex consists of layers with different cell distributions. These layers can be hierarchically organised, with the lowest level being closest to the sensory input and higher regions being further removed (Fig. 11.2).

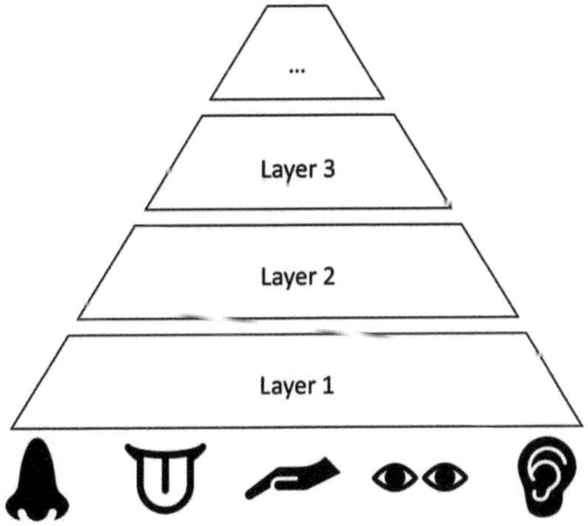

Fig. 11.2 Hierarchy of cortical regions where the first layer receives input from the senses

[16] A primer on PP is given by Wiese and Metzinger (2017). A scientific survey on the knowns and unknowns of the Bayesian brain is presented by Pouget et al. (2013).

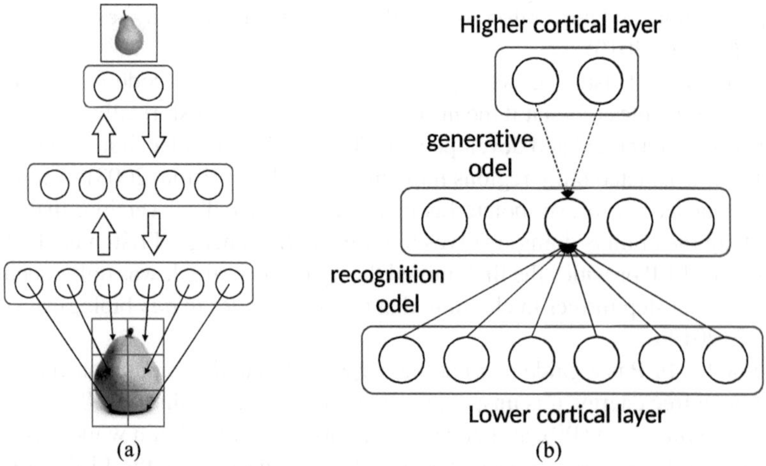

Fig. 11.3 Cortical levels interacting with each other. In (**a**) we see lower cortical regions having direct access to the sensory data, while higher cortical regions consist of abstract representations that predict the sensory input. In (**b**) we see that each neurone receives both top-down and bottom-up signals. Each neurone is bidirectionally connected to higher and lower layers, as indicated here for only the central neurone

PP models do not view the brain as passively representing the outside world, but rather as a system that is constantly engaged in trying to predict the present state of the world (Clark, 2013, 2015). The PP theory of the brain suggests that higher levels in a cortical hierarchy predict the activity of lower levels. When a mismatch in prediction is identified, there is a high level of surprise, and synaptic strengths are modulated (i.e. synaptic plasticity occurs), resulting in an update of the cortical model. Such an adjustment of the model is what we call *learning* (Clark, 2016; Friston et al., 2017; Hohwy, 2013). PP models describe mechanisms for minimising free energy and thereby minimising the long-term average prediction error.

According to PP, the different layers are thought to represent different levels of interpretation of the world. The brain is considered to be composed of predictive machines organised in a hierarchy of *abstraction*. At the bottom level, we find the raw sensory data that the agent acquires (see Fig. 11.3a). The first layer in this cortical hierarchy tries to predict the incoming data while removing noise from the signal by abstracting it to a more general representation.[17] The filtered signal is then passed upward to the next higher cortical layer, which again attempts to predict the incoming data, and so on. Each layer essentially uses top-down

[17] Even though it is the network as a whole that builds a representation of the world (see Williams 2018, for a detailed discussion on PP and representation), high-level layers that encode for high-level abstractions are here simply said to represent abstractions, without referring to the rest of the network.

Bayesian inference to achieve the goal of predicting the activity of lower layers and minimising information-theoretic surprise.

A computational model of the PP scheme applied to the visual cortex was first presented by Rao and Ballard (1999), and more sophisticated models have been developed ever since, taking more neurophysiological features into account. As a general overview, any PP model contains the following core elements (Wiese & Metzinger, 2017):

1. **Top-down processing:** Not only do lower-level cortical regions, closer to the sensory input, propagate processed information to higher cortical regions, but also vice versa.
2. **Statistical estimation:** From the data sampled over the sensory system, the brain builds a model that allows it to statistically estimate the probability of the current state of the world. Anything that is taken to be found in the world is the result of statistical inference.
3. **Hierarchical processing:** The estimates of the brain are hierarchically organised. Typically, regions in higher cortical areas will encode more complex and large-scale feature estimations.
4. **Prediction:** Higher cortical regions predict the activity of the lower regions. At the lowest level in the cortical hierarchy, the sensory input is predicted, whereas in higher regions neurones predict more abstract features.
5. **Prediction error minimisation and control:** A neurone tries to correctly predict the activity of neurones in lower cortical regions. If there is a mismatch in the prediction, the neuronal model is updated, typically by modulating synaptic strengths.
6. **Bayesian inference:** The procedure follows the norms of Bayesian inference.

Typically, the top-down information flow is realised by a *generative* process that models and samples from a probability distribution to generate predictions of network activity. The generative process is used at each network level to generate a prediction of the activity in that layer. As a result, the generative model can also be used to create possible sensory inputs (see Fig. 11.3b). This top-down processing is especially active when we imagine images in front of our eyes or when we dream. It is constantly engaged in generating imagery, typically for the purpose of making predictions, but also for fantasising about images one has not actually seen.

The information flow moving bottom-up from lower to higher cortical regions corresponds to a *discriminative* or *recognition* model that tries to correctly classify the input data, providing feedback to the higher regions as to whether a prediction made by the generative model was successful. If not, then the model recognises that it must adjust, as it has encountered a high level of surprise, which ought to be avoided.

PP presents an alternative theory to the standard, orthodox view of the brain. According to the traditional model, perception is a purely bottom-up information extraction process. The receptive fields of feature detectors are often thought to be hard-wired and innate to the brain. This view, however, has been challenged as untenable (Karni et al., 1994; Merzenich et al., 1983; Sharma et al., 2000). While the orthodox view focuses on a neural recognition model, where information flows upstream from the sensory input to higher cortical levels that encode higher abstractions, it neglects the crucial role of anticipation. Conversely, PP suggests that the higher-level abstractions are constantly involved in actively predicting lower-level activity. This is realised by a generative model that lets information flow in the opposite direction to the recognition model, thereby communicating the needs of higher cortical layers to the layers at lower levels (Fig. 11.2).

Three central beliefs about the brain were earlier identified in Quine's writings that characterise crucial assumptions made by the orthodox view: one-directional causal flow of information, predefined feature detectors, and categorisation by similarity comparisons. As we see, the first of these points is undermined by the two-directional process involved in perception. The second assumption also does not reflect itself in PP: since Quine's writings, the belief that the receptive fields of feature detectors are specified innately has been discarded by many neuroscientists and theoreticians working in machine learning (Hinton, 2007; Karni et al., 1994; Sharma et al., 2000). Perception is not a passive process in which the features of an object are detected by innate specialised modules of encoding. Instead, feature detection is constantly adapted to the system's needs in order to make good predictions. Higher cortical regions communicate to lower regions what kinds of features they ought to be sensitive to.

PP suggests that the brain is hierarchically structured and that higher cortical regions actively generate percepts, attempting to predict the sensory data in a robust way on the basis of prior probabilities and likelihood estimates. This is realised by a hierarchical network with a two-directional information flow, where the recognition network runs bottom-up from the incoming signals to the higher cortical regions, while the generative network runs top-down in the opposite direction. This structure allows lower cortical regions to send information about false predictions upstream and enables higher cortical layers to communicate the needed feature detection adjustments to the lower levels. The features to which neurones respond are not innately determined; rather, these features are continually re-evaluated and adapted, depending on the current requirements for making successful predictions (Clark, 2016; Friston, 2005; Hohwy, 2013; Kersten et al., 2004).[18] Contrary to the classical view, such as the one Quine relies on above, our visual experiences are shaped by the brain's need to make robust predictions about the activity in lower regions of the visual cortex. Whenever a prediction is inaccurate, the brain's internal model is updated; it learns and adapts through synaptic and possibly structural plasticity. According to PP, the problem of underdetermination is not solved by an innate, hard-wired feature detection machinery, but rather by a form of approximate Bayesian inference. Prior beliefs are encoded by generative networks and are used to generate

[18] Note that the notion of prediction is a technical one and refers to mechanisms of which we are not consciously aware.

percepts that predict the neural activations in lower cortical regions—ultimately the activity of those neurones that is directly caused by the sensory data.

To understand the distinction between the traditional view and that which is given by PP, let us compare the PP account to a system processing a feature like the motion of objects in a hard-wired kind of fashion—thus, in a way similar to what we have seen with the Reichardt detector circuit (Fig. 9.7 on p. 106). While in theory, the circuit only encodes the detection of very specific velocities v, we are dealing here with biological processes, and thus there is always a tolerance range in which the circuit is still active. The circuit is therefore active whenever the object's motion has a velocity in the range $[v_{\min}, v_{\max}]$. Now imagine two cognitive systems, A and B, repeatedly observing two objects that move at constant velocities of 2 and 4 m/s, respectively. We assume that A has feature-detection machinery of the kind of a Reichardt detector that can only discriminate between velocities in a range of 3 ± 1 m/s and in a range of 6 ± 1 m/s. System A would fail to discriminate between the objects because both would trigger only the first Reichardt detector circuit. Contrast this with the case in which we imagine cogniser B to discriminate objects by making good predictions using a PP scheme. Such a scheme can alter the features it detects, even if B also only had two feature detectors. The detecting circuits would, consequently, eventually adapt to discriminate the objects from one another. In this case, the feature detectors would, for example, become sensitive to velocities in the ranges of 2 ± 1 m/s and 4 ± 1 m/s, respectively. This form of discrimination increases the probability of making better predictions of future data because it improves the system's ability to track relevant behaviour in the data.

A cognitive system that does not rely on an innate feature detection machinery is adaptive to a continuously changing environment and can use its resources effectively. This is important when the complexity of the world is too great to be fully captured and requires the brain to compress sensory data to make *probabilistic* rather than determinate models of the world. Due to its limited resources, the brain cannot keep track of everything that is going on in the world but constructs an internal model of the world that is most *likely* to predict future events. For this, the model must be highly adaptive and cannot rely on innate feature detectors. Constructing relevant feature detectors, rather than having them innately engrained, is one of the essential techniques employed in PP to build an internal model of the world.

The paradigm shift from a passive to an actively involved brain is influenced by the overall developments and achievements of machine learning in the past two decades. Since its beginnings, machine learning has moved away from simple, hand-coded feature detection algorithms towards generative models. The generative model of a neural network creates imagined sensory data from a prior model such that the network can learn by self-supervision, that is, the input data correspond to the target data without external human supervision (or homunculus) being required.

11.6 Connecting the Personal and Sub-personal Level

Through the theory of PP, we understand the brain to be performing Bayesian inference by making predictions—minimising prediction error and thus surprisal—on the sensory input data, using an internal prior model of the world. The term *prediction* in the context of PP is closely related to our ordinary notion of making predictions, and thus the question arises how exactly these two terms relate to one another. In PP and the FEP, predictions occur at the *unconscious* level and are not the same as when we say that we are predicting the weather for tomorrow. The unconscious predictions discussed in the context of PP are predictions of neural activity. They are similar to conscious predictions in that both consist in probabilistic assertions about future events. However, in contrast to conscious predictions, unconscious predictions are not linguistic and mostly concern immediate forecasts of sensory input—although other unconscious predictions, such as the adaptation of muscle mass, are less immediate.

Similarly, the notion of surprise is ordinarily recognised through facial expressions rather than by information-theoretic considerations. For this reason, application conditions seem to vary, and some philosophers prefer to distinguish between *surprise* and *surprisal*, whereby only the latter refers to the information-theoretic notion. For the moment, let us adopt this distinction and use the term *surprisal* to refer solely to the technical concept that characterises neural plasticity, reserving *surprise* for the ordinary, manifest notion. A direct translation from surprise to surprisal is said to risk a *mereological fallacy*—the mistaken ascription of psychological predicates to parts of a person that are in fact only applicable to the person as a whole (Bennett & Hacker, 2003). According to Bennett and Hacker, psychological predicates apply only to whole humans (or non-human animals) and cannot intelligibly be applied to parts of the body, such as the brain. This is because psychological terms gain their meaning in the context of human behaviour. They claim that if someone is, for instance, thinking, then it is not the brain that is thinking but the person as a whole. This thought finds its origins in Wittgenstein, who asks:

> But isn't it absurd to say of a *body* that it has pain? ... If someone has a pain in his hand, then the hand does not say so (unless it writes it) and one does not comfort the hand, but the sufferer: one looks into his face. (Wittgenstein, 1953, §286)

Bennett and Hacker express their concern that the fallacy persists throughout large parts of the neuroscience literature.

I agree that what they call the mereological fallacy is, in many cases, a genuine fallacy found in the scientific literature, and I share their concern about these cases. The brain does not 'think', it does not 'feel pain', and it is not 'happy' or 'sad'. These predicates can only be attributed to the human being as a whole. They are highly disconnected from any neurological concept, and it is entirely unclear what their neurological counterparts are or whether they are realised in one or in a multiplicity of different ways. This is why their application conditions are only determined for humans as wholes, and we have no clear sense of how these terms could be applied to brains in isolation.

However, I disagree with Bennett and Hacker that the fallacy applies to *all* psychological predicates. Their argument is not successful in the case of *surprisal*, because of the concept's scale-invariant nature.

Surprisal is a concept we can apply to any information-processing system. The systems under discussion are systems that are bounded within a Markov blanket. As we have seen, individual neurones have a Markov blanket, the brain has a Markov blanket, and the body as a whole has a Markov blanket. In all these cases, surprisal is a well-defined term. What they have in common is that they exhibit high surprisal when prediction errors occur. According to the FEP and PP, the actions they perform are a consequence of this surprisal and an effort to reduce it in the long run. When we say that an individual cell has high surprisal or that a whole body has high surprisal, we are saying that the system is encountering an unexpected situation. How is this different from our ordinary notion of surprise?

One might think that our ordinary notion of surprise is mostly determined by how somebody looks—the raising of eyebrows and the widening of eyes. While these expressions are clearly related to the experience, they do not determine the meaning of the term. If somebody looks surprised, they are not necessarily surprised, as they might merely be acting. In such a case, we would rightly say that the actor is not *really* surprised. Looks are not essential to the meaning of being surprised, and looking surprised has a different meaning than being surprised. When we say that somebody is surprised, we are actually saying something about their inner state, not simply how they appear. The meaning of the term rests on an internal mental state of encountering an unexpected situation.

But this is the same definition as the formal definition of surprisal, with the only difference that surprisal is applicable to any Markov blanket whereas surprise is only applicable to humans and possibly animals. In what way, then, is the distinction useful? A person who is surprised also has high surprisal. And a person with high surprisal is also surprised. For humans, the two terms are essentially interchangeable, and the only difference is that the notion of surprisal is applicable to other systems as well, such as brains. But this again is exactly what the mereological fallacy tells us is a fallacy, which in this case seems to bear no problem at all. Therefore, we can just as well remove the term 'surprisal' from our vocabulary and accept that scientists have found out that 'surprise' is a much broader concept than what we had originally expected it to be. In the ongoing text, I will therefore only be speaking about surprise.

What we notice from these considerations is that PP bridges a gap between a manifest, psychological term and its scientific counterpart. This illustrates one of the central aspects of why PP is of particular relevance to us. PP can be understood as an *intermediate-level model.* That is to say, it is unspecific about precise neurophysiological details but aims to explain the most general computational methods that are in operation across different brain structures and different length scales (Spratling, 2013). It is especially relevant for our purposes as it bridges the gaps between understanding the general purpose of individual neural activity and first-person human experiences:

> The PP schema is especially attractive because it deeply illuminates the nesting of the neural economy within the much larger nexus of embodied, world-involving action. … PP suggests new ways of making sense of the form and structure of human experience …. (Clark, 2016, pp. 2–3)[19]

PP offers insights into the neural code that reflect themselves on all levels of investigation: at the level of cognitive and manifest phenomena down to the level of individual neural interactions.

Summary

Biological systems minimise the long-term average prediction error of their sensory inputs, given their previously collected sensory data. When we conceive the brain as being a system that avoids the organism's decay by regulation, the FEP becomes more than just an answer to Schrödinger's question on life but turns out to be a quantum leap in neuroscience research. The new perspective undermines classical models in cognitive science research and finds concrete realisations in PP models. These models are powerful as they relate the unconscious with the conscious, as exemplified by offering a case in which a psychological predicate is applicable to the brain in isolation without being prone to the mereological fallacy.

---

[19] Notice that this is not an uncontroversial issue. See, e.g., the exchange between Hohwy and Seth (2020) and Schlicht and Dolega (2021).

Chapter 12
Hyperpriors and Kant's Views on Cognition

In this chapter, I discuss how PP gives us a deeper understanding on the realism vs. idealism debate. We will see that much of PP largely points towards idealism and is in support of a debunking argument against realism. However, a conclusive argument will only be found in later chapters.

12.1 Experiencing the World as Controlled Hallucinations

Prior to the development of PP, some researchers have speculated based on early empirical findings and logical reasoning that the content of our experiences must to a large extent be a construction of our neural system. For instance, in the writings of the Nobel laureate Francis Crick, he recognises the computational difficulties involved in identifying objects and their categories from sensory input. The active nature of the process required to develop a symbolic interpretation of the environment, makes him suggest that the brain impose its own symbolic representation of the world onto the data:

> It is difficult for most people to accept that what they see is a symbolic representation of the world – it all seems so like 'the real thing'. But in fact we have no direct knowledge of objects in the world. This is an illusion produced by the very efficiency of the system. (Crick 1994, p. 33)

The biologists Maturana and Varela (1984) also make strong claims about our manifest experiences being neural constructs. Just like Hoffman, they subscribe to the belief that for an organism to be successful in evolutionary terms, its neural system converts sensory data into representations that are particularly useful for survival but are largely not veridical. They argue that reality is not independent from the observer but rather underlies the observer's neural construction: '[We must see] cognition not as a representation of the world "out there," but rather as an ongoing bringing forth of a world through the process of living itself' (Maturana & Varela,

A. C. Schwaninger, *Cognitive Metaphysics*, Synthese Library 529,
https://doi.org/10.1007/978-3-032-05420-3_12

Fig. 12.1 Blind spot self-test

1984, p. 11). This claim is initially motivated by two phenomena that are easily self-verified. First, consider the well-known case of so-called *blind spot*. The region on our eye's retina through which the optic nerve passes, is a region that lacks light-detecting photoreceptor cells. Thus, the corresponding region of the visual field is 'blind.' Ordinarily, we do not notice that our visual field is obstructed but it can be observed by a simple test on oneself: Close your left eye and focus at a normal reading distance with the right eye on the left circle in Fig. 12.1. Now slowly move the image closer to your face without losing focus of the left-hand circle. At a certain distance, the other circle will appear to have vanished.

What is fascinating about this phenomenon is that the cognitive system does not let us experience a black hole in our visual system. Rather, it fills the void with a surrogate of its own imagination in such a way that we are generally not aware of our visual deficiency. With regard to the region of the blind spot in the visual field, what we see is not what is 'out there' but what is simulated by the brain.

This is further illustrated by the second example of so-called *coloured shadows*. In 1672, the German scientist and politician Otto von Guericke noticed that the shadow of his hand appeared bluish while being exposed to both candlelight from one side and sunlight from the other. He found that when red light hits an object while white light is shining from a different direction, the shadow created appears to be blue. While scientists speculated about the nature of this phenomenon, it was later clearly shown that the light emitted from the shadow region did not contain wavelengths that we associate with the colour blue. Instead, the bluish colour tone we see is a construction of our cognitive system. While some might find the bluish colour to be an illusion in contrast to what the colour *really* is, Maturana and Varela suggest that

> [we must] stop thinking that the colour of the objects we see is determined by the features of the light we receive from the objects. Rather, we must concentrate on understanding that the experience of a colour corresponds to a specific pattern of states of activity in the nervous system which its structure determines. . . . [W]e can demonstrate that because these states of neural activity . . . can be triggered by a number of different light perturbations . . ., we can correlate our naming of colours with states of neural activity but not with wavelengths. . . . [W]hat we took as a simple apprehension of something [such as colour] has the indelible mark of our own structure. (Maturana & Varela, 1984, p. 22)

The findings are interpreted to challenge the metaphysical realist's commitments about colours as properties of things, existing independently of the observing mind. The brain does not just passively extract information from the sensory data it receives but is actively adding information to it. As some would say, 'the world we experience is really a neural simulation' (Varela et al., 1991, p. xix), a neural construct that is mistaken for reality itself (Metzinger, 2009, pp. 6–7).

Turning now towards the recent development of PP, we find that these novel models explicitly underline the active nature of the brain. As Dan Zahavi (2018) explains, PP supports the view that the reality we perceive is largely a construct, due to its active nature of top-down information processing. He explains his views as follows: 'the claim isn't simply that our access to the external world is mediated by neural representations, but rather that the world of experience is itself a representational construct' (p. 48). This does not mean that there exists no external world, but that all the structuring of the experienced world is a result of percepts being generated by higher cortical strata and imposed on lower strata in the cortex. This is what some have referred to as a form of 'controlled hallucinations':

> [A] fruitful way of looking at the human brain ... is as a system which, even in ordinary waking states, constantly hallucinates at the world, as a system that constantly lets its internal autonomous simulational dynamics collide with the ongoing flow of sensory input, vigorously dreaming at the world and thereby generating the content of phenomenal experience. (Metzinger 2004[2003], p. 52)

12.2 Alignment with Kant's Views on Cognition

Our traditional view of the brain as a passive, stimulus-driven device for processing sensory data is replaced by a more modern theory suggesting that the brain actively performs top-down control from higher to lower cortical regions (Engel et al., 2001, p. 704). This idea has been claimed to align with and be rooted in Kantian idealism (Swanson, 2016). Prominent philosophers of the cognitive sciences have noted that PP seems highly reminiscent of Kant's understanding of cognition.[1] PP has been labelled a 'neo-neo-Kantian view of the relationship between mind and world' (Anderson & Chemero, 2013, p. 204), and it is said that 'there is certainly a distinct Kantian element' (Hohwy, 2013, p. 5) to PP or that it evokes an 'almost Kantian feel' (Clark, 2013, p. 196).

Contrary to classical neuroscience, Kant's understanding of cognition comes from reversing the relationship between cognition and objects, granting cognition constitutive power. For Kant, it is a futile enterprise to investigate the world *in itself* without turning towards our cognitive faculties. He attributes constitutive power to our cognition and says 'that objects must conform to our cognition' (B xvi), since 'imagination is a necessary ingredient of perception' (A 120). This view on cognition is in line with PP, according to which the sensory data does not determine how we carve up the world of our experiences, nor does it trigger in us a determinate response, nor is our perceptual apparatus adapted to detect predetermined features

[1] For a good overview of some of the parallels between PP and Kantian philosophy, see Swanson (2016).

in the world. Instead, the cognitive system detects features in the data which it finds useful at a certain moment in time in order to make robust predictions about upcoming sensory input using existing categories. Our visual experiences are generated by the brain imposing certain categories onto the data which are expected to minimise prediction errors. It is not that the categories are passively extracted from the data.

While both Quine and Kant would agree that the problem of underdetermination exists, and that sensory data is insufficient to determine what lies 'behind' it, their views differ in what the problem implies and in its philosophical significance. According to Quine, sensory data generates a certain response in us because of our innate feature detection machinery. Science will then confirm our experiences by providing coherent models of reality.

What Kant would say about the problem of underdetermination can be inferred from his account of how non-empirical concepts are connected to sensory stimulation. The condition that allows for the experience of objects 'is itself, at all times only a product of the imagination' (A 140). When we see, for instance, a triangle, '[n]o image whatever of a triangle would ever be adequate to the concept of a triangle as such. For it would never reach the concept's universality that makes the concept hold for all triangles, but would always be limited to only part of this set' (A 141). Thus, to see a triangle requires us to impose the concept onto the data. Consequently, we do not merely decode what lies behind the sensory data, but we impose our concepts upon it.

This perspective of Kant seems to anticipate the idea of the brain generating images for predictive purposes. If we look again at the Dalmatian dog image (Fig. 9.5 on p. 104), the cognition process becomes more apparent. It is not that the brain extracts features from the image in order to find the object encoded in the data; rather, it uses the information received to select a class it already possesses and carves up the image accordingly. This undermines Quine's conviction that categories are assigned to objects on the basis of their similarity to other objects and how we have learned to speak about them. Instead, according to the PP perspective, we already possess the categories within us and impose or re-identify them in the data. We have the category 'dog' and impose it onto the image such that we in fact see a dog. However, according to PP, the predictions we make can vary depending on experience and needs. The hidden causes of our sensory perceptions are thus not the consequence of pre-individuated objects.

This process will be discussed in more detail in the next chapter. For now, it is important to understand that the first idea reminiscent of Kant is that we cognise the world through an active process, that is, PP reverses the relationship between cognition and object. This occurs through the generation of percepts aimed at predicting the activity of lower-level neurones triggered by sensory data. The second idea is that we possess innate representations—so-called *hyperpriors*—of how the world is, which determine all our experiences by presupposing what the world of our experiences should be like in order to make successful predictions. Let us now turn to this second point in more detail.

12.3 Hyperpriors in the Brain

We have encountered Bayes' Theorem as a central aspect of neural processing, which can be understood as a rule for updating the brain's internal model in light of new information. So far, we have focused on how *a posteriori* knowledge is formed. However, the rule also depends on the availability of a prior probabilistic model. But where does this initial prior model come from?

To some extent, priors must be innate to our cognitive framework. Ultimately, there must be some prior at which our reasoning starts that comes from how our neural system is encoded prior to any experience. These priors are what we call 'hyperpriors' (Clark, 2013; Friston, 2013a; Hohwy et al., 2008). They are initial biases that direct our perception and experiences into a certain direction. From all possible experiences we could have, based on some given sensory input, hyperpriors impose a certain structure onto the data that narrows down all possible structures we could experience to just a small number of structures we really experience. Typical examples of information encoded in a hyperprior would be 'that there is only one object (one cause of sensory input) in one place, at a given scale, at a given moment' or that 'we can only perform one action at a time, choosing the left turn or the right but never both at once' (Clark, 2013, p. 196). Stating that there *is* only one object at a place at the same time is the realist way of formulating the examples, suggesting that it is physically impossible for objects to coincide and that our hyperpriors have adapted to this physical fact. The idealist would, of course, simply say that some hyperprior is more useful than another for our survival without making claims about how the world really is. Hyperpriors form a selection of structures that are useful and assign a zero probability to those structures that are not worth being further considered.

Within the PP framework, it is suggested that hyperpriors are present at each level of the hierarchical neural system (Clark, 2013; Kemp et al., 2007, p. 2007; Tenenbaum et al., 2011), whereby the higher we are in the system, the more abstract the assumptions encoded in the priors are. The connection to Kant is evident, as he assumes that there must be some predefined principles within our perception that are not based on empirical observation but can only be revealed by transcendental arguments. Andy Clark mentions this connection in passing: 'Hyperpriors are essentially "priors upon priors" embodying systemic expectations concerning very abstract (at times almost "Kantian") features of the world' (Clark, 2015, p. 174).

There is, however, no unique connection between PP and hyperpriors; in fact, any learning algorithm requires some prior on what should be learned. Otherwise, learning is not possible. According to statistical learning theory, 'the incorporation of prior knowledge, biasing the learning process, is inevitable for the success of learning algorithms' (Shalev-Shwartz & Ben-David, 2014, p. 3). In other words, all learning involves prior knowledge about the system it is intending to learn something about, which means that the learning mechanisms of the cogniser must already be sensitive to detecting the features of a pattern in order to detect the pattern.[2]

To illustrate this, consider a study on the behaviour of rodents by John Garcia and Koelling (1996). What they demonstrated empirically was that not everything can be learned, and thus there are cases where an organism will never be able to develop an adequate cognitive process for a certain situation or problem it is facing. In a variety of experiments, rats in cages were either electrically shocked, exposed to loud noise, or subjected to a bad odour whenever they were eating from some specific food in the cage. What one would ordinarily expect is that the rats would eventually be conditioned to learn the correlation between stimulus and food and thereby avoid the one food source in favour of another. In this experiment, however, the associative learning process did not take place. The rats were unable to be conditioned on the presence of food while being exposed to an unpleasant stimulus.

These results can be interpreted to show that, over the course of evolution, an organism obtains a priori beliefs about its environment that bias the learning mechanism and might not allow the organism to learn a certain response or behaviour to a given problem. This is reflected in the PP scheme by a prior probability distribution that depends on hyperparameters which cannot be altered by the organism itself but only through genetic variations in the species' evolutionary development. In that case, we must interpret that the rats possess some unconscious a priori belief[3] that electric shocks cannot be caused by food and that noise does not actually relate to the quality of food. This form of an a priori belief is not adaptive and restricts the organism from accessing alternative hypotheses, even if they would be useful. The organism is hindered in learning the correlation in question and is not able to form any new beliefs that stand in contrast to the a priori belief.

One might be inclined to think that this is only true for some organisms with a cognitive system that does not compare to the human's. When we look at our own learning capabilities, it seems that there must also be a *universal learner*—a computational model that can be applied to any non-random data and will successfully learn the features and therefore a representation for this data. This, however, has been proven to be false mathematically.

It can be shown that every learning algorithm makes prior assumptions about the problem at hand. This fact in learning theory has the consequence that there is no universal learning algorithm, as expressed by the *No Free Lunch Theorem* (Wolpert & Macready, 1997). According to this theorem, there is no universal learning method, since the average performance of any two learning methods will be the same across all possible problem instances drawn from a uniform probability distribution. If the opposite were the case, namely that there existed a universal learner capable of learning any problem purely by statistical reasoning, then the

---

[2] Notice that the notion of knowledge is not understood in a propositional sense as in 'knowing that $P(x)$' but rather in an information-theoretic sense.

[3] Some might prefer to refer to this as a *cognitive propensity* rather than a belief. What follows should be true for either formulation.

algorithm would be completely unbiased towards the structure of the dataset. But finding such an algorithm is impossible, and therefore prior knowledge about the problem at hand is always required. Intuitively speaking, since all learning can in principle be expressed in terms of Bayesian inference, which presupposes some prior probability distribution, learning cannot occur without prior assumptions. There cannot be a learning algorithm that does not carry any prior information about the system it is intended to learn. Therefore, any organism will fail to appropriately learn and respond to some highly unusual situation in its environment.

12.4 Biased Hyperpriors

While hyperpriors are essential, the question arises whether those encoded by our brain structure are in correspondence with the world or whether they are biased away from it. In the 1950s and 60s, the statistician Charles Stein made a striking discovery. Intuitively, one would expect that whenever we construct a scientific model from some data, there will always be an unbiased model that outperforms any biased model.[4] This would support the intuition that evolutionary processes should shape humans to perceive the world in an unbiased fashion, resulting in non-biased hyperpriors.

Together with his student, Stein showed that this intuition is false. In other words, they demonstrated that for some given data, it is by no means true that an unbiased estimation of a probability distribution is always the optimal estimation (James & Stein, 1961). For instance, imagine one collects samples from a multivariate Gaussian probability distribution in more than two dimensions and wants to estimate the Gaussian's mean. Stein proved mathematically that there is a biased estimator that is systematically better, in the sense of quadratic deviation, than the mean of all the samples (Stein, 1956).[5]

This strikingly counter-intuitive conclusion can be expressed more generally in terms of the so-called *bias-variance trade-off* (see, e.g., Geman et al., 1992). According to this principle, there is a trade-off between the bias and the variance of a model. This is a trade-off between trusting in one's prior beliefs and adapting one's model to the data. It might, for instance, be possible that a model with a high bias due to strong prior beliefs is better than one with low bias, because it has a substantially smaller variance (i.e., the model has a low deviation from the true value when exposed to novel data).

[4] The bias of an estimator or model is the difference between the expected value and the true value of the estimated parameter. It quantifies how 'close' an estimator is to the value being modelled. In contrast, the variance of an estimator measures the deviation of the estimator when trained on or exposed to different data.

[5] The mean of all samples is in this case the maximum-likelihood estimator.

In the same way as computational models are subject to the bias-variance trade-off, so is our mind subject to the same effects. We might be able to make highly successful predictions of our sensory data despite being biased. Feldman (2013) points out that our cognitive system might well be biased away from a most accurate representation of the data. He suggests that Bayesian priors in cognitive science models are often falsely interpreted to be true reflections of occurrences in the world. Instead, he shows that our cognitive priors are most likely biased and claims that this supports those who suggest that human and animal cognition might not be directed towards truth (Hoffman, 2019; Hoffman et al., 2015; Mark et al., 2010).

This indicates that there is little reason to assume that our priors are fitted to the world. According to realism, there is an isomorphic relationship between the Px-structure of our thoughts and the structure of the world. But as we have seen, some cognitive scientists believe that PP undermines realism due to its active nature of imposing structures in a generative, top-down fashion. Many of the structures of our visual experiences are not structures that the brain extracts from the data but rather structures resulting from the brain's 'controlled hallucinations', imposed onto the sensory data based on earlier experiences. Hyperpriors determine the most abstract principles of our cognition, and learning theory suggests that there are no strong theoretical reasons why these priors ought to be in line with how the world actually is. In the next chapter, I will explicitly show how our visual structures can be derived from the principles of PP to debunk realism.

Summary

Much of PP supports a version of Kantian idealism. Not only does the brain's active nature, involving top-down processing, undermine many realist assumptions, but there are also reasons to believe that hyperpriors are innately biased, shaping our experiences in ways that are not necessarily veridical. Studies show that hyperpriors are very likely to encode assumptions about the world that are biased away from how the world is actually structured, in favour of more useful assumptions that enhance the organism's chances of survival.

Part III
Constructing the Manifest World

Chapter 13
Individuation by Prediction and Debunking Realism

In this chapter, I explore how visual objects are individuated from the background through the framework of PP, giving rise to the experience of a Px-structured visual ontology. This explanation does not hinge on any presupposed structural assumptions about the external world but emerges purely from the workings of neural processes. Consequently, this reinforces the idealist view that the Px-structure is imposed by the subject onto sensory data, offering a debunking argument against realism.

13.1 Learning a Representation Requires Having a Representation

Early accounts of perception tend to treat object detection and categorisation as two separate issues. Seeing an object x of kind F is thereby a two-step process of first detecting objects in the visual data and then categorising them. We have already encountered Wertheimer's (1923) proposal concerning the first step of segmenting objects from the background. According to him, there exists a cognitive mechanism that groups together 'elements' of the visual field, resulting in the perception of discrete, individual objects. As we have seen, Palmer and Rock (1994) provided an alternative grouping principle called the uniform connectedness principle, which closely resembles the information-theoretic approach. In either case, only once the object's boundaries have been identified such that it can be distinguished from its background, does categorisation or classification take place.[1]

This chapter is derived in part from an article *What Simulations Teach Us About Ordinary Objects* by Arthur C. Schwaninger, taken from *Open Philosophy* ©2019 The Author. Published by Walter de Gruyter.

[1] The terms 'categorisation' and 'classification' are used interchangeably in the context of PP.

A. C. Schwaninger, *Cognitive Metaphysics*, Synthese Library 529,
https://doi.org/10.1007/978-3-032-05420-3_13

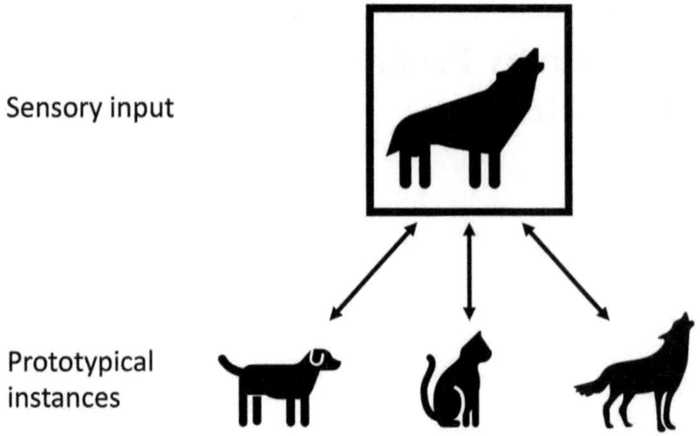

Fig. 13.1 Naïve object classification

To account for how categorisation takes place, many early explanations assume, in one way or another, a static similarity comparison between the visual data of the detected object and prototypical instances of categories stored in memory. This view is illustrated in Fig. 13.1, where the sensory input at the top is compared to prototypical instances of the classes 'dog', 'cat', and 'wolf' at the bottom. According to this *naïve view on classification*, the brain determines the sensory input's category based on its similarity to the prototypical instances.

The naïve view suggests that an object's category is determined by the brain computing some '"comparisons" of a presented [visual] patch with some sort of standard or prototype in memory' (Raffman, 1994, p. 48) in order to evaluate the category F of an object x. Thus, if the perceived object x is similar to an image of type F, or to a set of well-defined, static features associated with a type F, then x is also of type F. This intuition is reflected in early computational models of object recognition, in which a large number of categorised images are stored in memory and compared to the sensory input image (see, e.g., Abu-Mostafa & Psaltis, 1987; Hopfield, 1982; Huberman & Hogg, 1984; Kohonen, 1978; Willshaw et al., 1969). Many theorists of perception have, however, rejected the idea of the brain storing a large number of picture-like representations in order to determine an object's category (see, e.g., Biederman, 1987; Ullman, 1989).

As we have seen earlier, both the uniform connectedness principle and the information-theoretic approach suffer from the problem of explaining how disconnected regions are individuated as a single object, and how regions of similar appearance can be experienced as distinct objects of different kinds. The major revelation of PP is that it resolves these problems by challenging the assumption that individuation and categorisation occur separately. Instead, PP relies on the idea that the features by which objects are individuated and distinguished from the background (as determined by the recognition network) are defined by their categorical membership (as determined by the generative network).

This idea may at first seem paradoxical: one would expect that, in order to learn a category, it is necessary to have already detected objects with similar features. Thus, how can categories serve as a criterion for object detection? The somewhat nativist idea that prior representations of the data (in terms of a probability distribution function) are required to train the recognition model seems disconcerting. If the neural recognition model depends on an internal, *a posteriori* model of the world, there appears to be a chicken-and-egg problem. How does the internal model, which qualifies as a representation of the world, come about in the first place? We seem to be confronted with an informational circularity.

Geoffrey Hinton (2007), one of the pioneers of deep learning, discovered an algorithm that makes it possible to detect and categorise an object simultaneously by treating individuation and categorisation as a single process of tiny, incremental representational improvements in a continuous space. This algorithm, which was inspirational for PP, shows how representation and individuation do not occur separately, and explains the close connection between object and predicate. To understand this connection better, let us look at some of the details.

The solution to the paradox is found by running and analysing simulations that show how the representation of the network units emerges spontaneously over the course of the learning process. The chicken-and-egg problem arises only from within our folk-psychological or classical notion of 'representation' as something absolute and binary (see, e.g., Fodor, 1998). From a machine learning and computational neuroscience perspective, however, there is no question of whether a cogniser has a representation or not, as it always has *some* representation—even if that representation is very poor. Thus, instead of thinking in terms of a cogniser either having or lacking a representation, we must consider a representation to be either *better* or *worse*, given the cogniser's innate cost function.

Together with others, Hinton was among the first to discover this new method of learning (Hinton et al., 1995; Hinton, 2007). He explains how the problem of circularity is resolved technically in artificial neural networks by giving the system an additional temporal dimension, where it alternates between different state phases:

> We have a chicken-and-egg problem: Given the generative weights we can learn the recognition weights and given the recognition weights we can learn the generative weights. It turns out that we can learn both sets of weights by starting with small random values and alternating between two phases of learning. In the 'wake' phase, the recognition weights are used to drive the units bottom-up, and the binary states of units in adjacent layers can then be used to train the generative weights. In the 'sleep' phase, the top-down generative connections are used to drive the network, so it produces fantasies from its generative model. The binary states of units in adjacent layers can then be used to learn the bottom-up recognition connections. (Hinton, 2007)

The chicken-and-egg problem is solved by the network model performing small, iterative steps, alternating between the waking and sleeping phases, thereby allowing the generative model to predict the input. Even though the cogniser's neural network starts off with a poor representation, in each iteration step towards finding the global minimum in the space of all possible models, the quality of the representation improves.[2] Thus, any arbitrary internal model can count as a representation, even if it is a very bad one. The representational model will still incrementally improve—essentially independently of the initial conditions[3]—until a better model is found. The wake-sleep algorithm, which was inspirational for PP, is a computational procedure showing how an optimal representation can be derived from an existing representation. This algorithm enabled the development of PP schemes that function in an unsupervised manner—that is, without requiring any human input regarding the categories of the world.

In PP, the higher cortical abstractions are non-*linguaformal* representations that have been adapted to make more successful predictions about incoming sensory data in the future. From the idealist perspective, the representations should clearly not be interpreted as representing things in the world. They are merely abstractions for the purpose of making successful predictions. The higher we move up the cortical hierarchy, the more robust these representations become, as they encode slower-moving regularities within the data (Friston, 2008; Hohwy, 2013, p. 27). While low-level representations encode what we ordinarily conceive of as *percepts*, high-level representations encode longer-term expectations on the data that correspond to what we think of as *concepts* (Hohwy, 2013, p. 72). Together, they allow us to 'carve up the world' in the way that is most useful to our needs. The particular concepts that a cognising system perceives shed light on the specific patterns in the sensory data that are relevant for making good predictions.

Representations can, to some extent, be visualised in terms of different domains within a latent *representation space*, where each dimension corresponds to the activity of a single neurone in the neurone layer that spans the representation space (Fig. 13.2). Here, we are only interested in the representation space of higher-level cortical neurones, where the activity in one of these domains results in the experience of a general abstraction or property in the data, depending on whatever concept the region encodes. If the sensory input (or a part of it) results in the activity of a domain, and a different input triggers the same domain, then the inputs are perceived to share a common property, such as there being a cat in the input when the domain encodes for abstractions we typically refer to as cats.[4]

---

[2] Computational Bayesian inference is generally conceived as an optimisation procedure that seeks to optimise some functional by means of variational calculus. The wake-sleep multistep iteration procedure can itself be understood as a function that is approximated by the neural network to iteratively find an optimal model w^* (the global minimum in the space of all possible models).

[3] This means only that a generic initialisation case within some neighbourhood of the solution will succeed in producing a useful representational model. It does not mean that *any* initial condition will lead to an optimal w^*. For instance, assigning the zero vector as an initial condition has been shown to perform much worse than random initialisation. In the brain, the space of possible values for w is already constrained by what is neurologically feasible, such as the range of synaptic strengths or the maximum neurone firing rate.

[4] Similarly, Gärdenfors (1996, 2000) suggests employing the notion of conceptual spaces that allow concepts to be represented geometrically by means of metric spaces. This leads to the proposal that identity relations in the manifest image are ambiguous and that it is from this quality that paradoxes such as the Ship of Theseus arise (Douven and Decock, 2010).

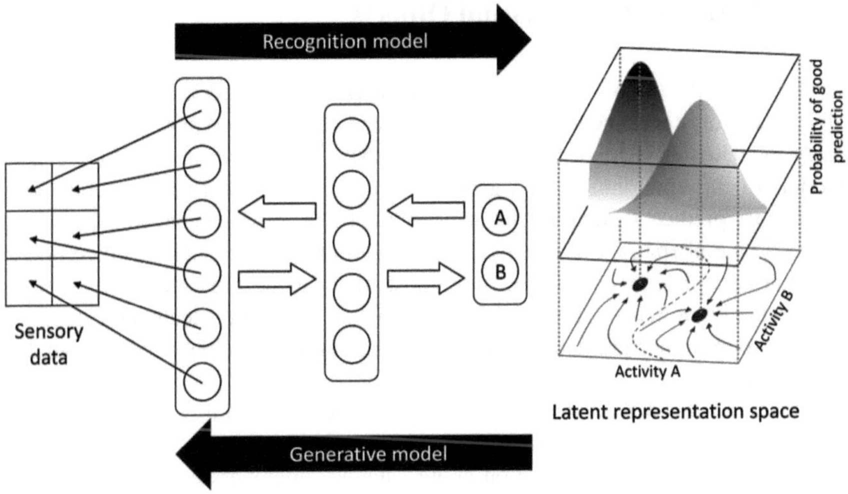

Fig. 13.2 Higher cortical layers build abstract representations of lower cortical layers. The network's upstream information flow is determined by the recognition model, mapping the sensory data onto the latent representation space. The network's downstream model is the generative model that creates images and tries to 'guess' the activity of lower levels and eventually the sensory input. The activity of the two highest cortical neurones, A and B, spans a latent representation space of the abstractions in the sensory data. When a representation is found, some given sensory data results in the activity of the most probable region of the representation space (illustrated by two peaked probability distributions). This activity converges towards the attractor (black circles)

The activity of the higher-level neurones will typically converge towards some *attractor* point, which is the representation's prototypical representation point. These aspects of PP are illustrated in Fig. 13.2, where we see a neural network exposed to sensory data that is represented in a latent, higher cortical representation space. Since there are only two neurones at the highest cortical level of the figure, the representational space will consist of only two dimensions. A more rigorous approach to representation or configuration spaces can be found in many standard texts on neural networks (e.g., Hertz, 1991).

The striking idea of the computational scheme behind PP is that higher-level feature detectors can make demands on lower levels to satisfy their needs. From a Bayesian perspective, the network contains prior knowledge that is used to update low-level representations through a downstream information flow. High-level representations enable the generation of *imagined* low-level neural activity patterns (such as imagined visual images) that can be compared with the *actual* lower-level activity. In machine learning terms, the top-down generative model creates imagined training data for the bottom-up recognition model in order to train the recognition model's weights. In addition to sensory data, top-down generated data are created for the sake of the network's own learning.

13.2 Individuation of Visual Objects

In the case of the brain's cerebral cortex, 'predictor neurons code information about object category; error neurons signal mismatches in predicted and observed object category' (Koster-Hale & Saxe, 2013, p. 838). Higher cortical regions encode models of the environment that generate hypotheses in the form of perceptions, aiming to predict structural characteristics of the experienced world such as categoric membership. For instance, given a certain stimulus, the cognitive system computes the category F that it finds most likely to be present in the stimulus. Conceptually similar to Hinton's algorithm, when the brain selects the category F in order to minimise prediction error, it also determines the object and its boundary in the stimulus.

This can be generalised to any cognising system with finite resources and the objective of minimising prediction error. Such a system must have some representation that encodes information about categoric membership, because sensory data are only predictable when considered as instances of categories that generalise certain features in the data for compression purposes.[5] A system can roughly be said to *impose* the mental category F, learned from earlier experiences, onto images. Imposing a category onto an image successfully (i.e., with little prediction error) allows the system to individuate objects in the visual field, for instance, by computing for all parts in the image the probabilities for belonging to an instance of a certain mental category.[6] Not imposing, or being unable to impose, any category leads to a maximally high prediction error since no visual object is individuated for the purpose of tracking it and predicting its behaviour. For a cogniser A to minimise prediction errors, it is therefore essential to individuate objects in the visual image provided via the visual system:

Principle of Individuation A region in the visual image is individuated and tracked as an object x if this robustly maximises the probability of minimising A's average prediction error.

The *visual image* refers to the information received at the retina and structured in the form of a two-dimensional grid corresponding to the physical arrangement of

[5] Despite our ordinary impression of there being a qualitative difference between concepts as general mental categories and percepts, this difference is only gradual in PP: 'Percepts are ... basically shorter-term expectations and concepts longer-term expectations' (Hohwy, 2013, p. 72).

[6] As Rescher correctly points out, concepts or general mental categories are required for individuation and identification: 'The fact is that ideation is an indispensable preliminary and requisite for individuation ... To be an identifiable item, that is to say, is to answer to the idea of a certain sort: identification is not possible without sortification and sortification is not possible without recourse to the corresponding ideas' (Rescher, 2016, p. 10). Although some scholars avoid the term 'category' in favour of 'representation' or 'hypothesis', I continue to use the term 'category' as it is in line with some of the Neo-Kantian literature and reminds us of the bridging character of PP between the personal and sub-personal levels.

Fig. 13.3 Image is adapted from Clark (2015)

the photoreceptors. From the retina, this information is passed forward to the lateral geniculate nucleus and the visual cortex.

If we turn back to Fig. 9.5 (p. 104) of the, at first glance unrecognisable, Dalmatian dog, we find that it illustrates how the theoretical insights of PP about perception manifest themselves at the level of our conscious experiences. The category of the dog is imposed onto the image because this is the assumption with the highest probability of resulting in small prediction errors when tracking the spots in the image over time. Without already possessing the right category—in the sense of being the most useful—it is impossible to detect the dog in the image, and methods like those proposed by Wertheimer or Palmer and Rock would clearly fail.

How prior perceptions affect the outcome of the category imposed onto the sensory data can be illustrated by looking at Fig. 13.3. When our eyes move from top to bottom, we first see the letter 'A', followed by the letter 'B', since the latter is the most probable category to be displayed after the former, given the sensory input. For analogous reasons, when we move our eyes from left to right, we see the shape in the centre as the number '13'. Whatever category is imposed, the corresponding regions of the visual field are then treated as composing a single object. The decision about which category ought to be imposed is made on the basis of prior experiences.

The PP framework opposes the naïve view of categorisation by suggesting that minimising prediction error is what determines categorisation and, with this, the individuation of visual objects. Not only does PP reject the idea of the brain making similarity comparisons between features in the sensory input and some *fixed* pictorial representation in the brain, it also suggests that there is no similarity comparison to statically predetermined features. Rather, the features against which the sensory input is evaluated are always evolving and changing.

This does not mean that, according to PP, the cognitive system makes no similarity comparisons. When the cognising system generates its predictions of the input, it relies on the generative network that models the joint distribution over both the input and the approximated 'hidden causes', which allows it to generate instances of predicted input data. The cognitive system learns and adapts according to the similarity between this joint distribution and the distribution over causes given the input data as expressed by the recognition model (Kiefer & Hohwy, 2020).[7] In contrast to the naïve view on cognition, which imagines some similarity comparison between external features and fixed, possibly innate features stored in the brain, the generative network encodes for changing feature-detecting structures. The brain makes a comparison between the sensory data and the self-generated predicting image. This comparison is context dependent and not static, but is under constant change in accordance with the evolving features that define the respective category. When the self-generated image is a poor prediction, the neural network model adapts in order to achieve better prediction results in the future.

13.3 Accepting Idealism

We have seen that top-down processing involves imposing structure on data, and we have seen that hyperpriors cannot be trusted to reflect structures in the world. We have also seen that compression is cogniser-relative and that patterns cannot be distinguished by objective means. All these points place pressure on realism, but the strongest argument against realism comes from a classical debunking argument. Given the foregoing reflections and the *Principle of Individuation*, we now see that we have an explanation for why objects and their properties are individuated in subject–predicate form that is independent of the structure of the world: the brain selects a category that acts as a generalisation for making successful predictions. The category is associated with features and determines what objects we extract from the data in an at first seemingly paradoxical manner. Since living systems are less complex than their environment and since they aim to sustain themselves, they can only do so by having categories as generalising compressions which they impose on the data to make predictions.[8]

This predicting process, which starts with the imposed category rather than the object itself, must result in a Px-structure as the system tries to detect data points that match the predictive assumptions by imposing the category *P*. This explanation for why we experience the world as individuals in a Px-structure does not in any way assume that the world in itself has this structure. Thus, unless it is merely by coincidence, the world and the data we access have no Px-structure. People suffering from visual agnosia illustrate this point by having access to all data without access to a structure of this data that allows them to reason about what they see. PP offers a full explanation of how data receives its structure without presupposing that the

[7] This similarity between distributions can be quantified in terms of the Kullback–Leibler divergence (Cover and Thomas 2006, ch. 2.3).

[8] We clearly do make some mind-independent ontological assumptions about the world, but these are about the world's complexity as a whole, not about the Px-structure. The purpose is to debunk the realist's assumption of a mind-independent Px-structure.

mind-independent world is composed of separable objects with certain qualities. It can thus be said that with PP we have debunked realism.

Despite many having suggested that 'the mind-independence of objects actually shows up in sensory experience' (Campbell & Cassam, 2014, p. 165) and having spoken of 'the fact that our experience does manifestly present a mind-independent world' (Martin, 2006, p. 406), the experiences we have of a world consisting of individuated objects reflect our innate neurological structures, just as the green colour of the houses in Emerald City reflects the green-tinted spectacles worn by Dorothy in *The Wonderful Wizard of Oz*.[9]

To understand the resulting idealist position better, let us examine two typical concerns people have with idealism and provide answers from the perspective I am offering. For some, it traditionally appears that the anti-realist or idealist position is either incoherent or simply banal in its transition from descriptions about the world to metaphysics itself. As Simon Blackburn writes:

> Idealists always face the problem of finding an acceptable way of putting what they want to say about the involvement of the mind in the world. Some fudge it: it is quite common to find people writing that 'objects' do not exist outside of our conceptual schemes, or that we 'create' objects (values, numbers) rather than discover them. This is not a good way to put anything. With the inverted commas off, such remarks are false. (We do not create trees and galaxies, nor the wrongness of cruelty, nor the evenness of the number 2. Nor can we destroy them either, except perhaps for the trees.) But what can they mean otherwise: what is meant by saying that 'trees' are mind-dependent, if trees are not? Perhaps just the platitude that if we did not have minds of a certain kind, we could not possess the concept of a tree. The problem for the idealist, or the anti-realist in general, is to steer a course between the platitude and the paradox. (Blackburn, 1984, pp. 218–219)

Of course, there is matter in the mind-independent world. (I avoid saying 'hunks' of matter in the plural, since any plurality again assumes individuated objects.) But as we have seen, the idea that there is a symbol 'x' in our heads that refers to an x in the world is a simply untenable view to hold. To evaluate the mind–world relationship in order to make progress on what the idealist might mean by saying that the world is mind-dependent, an analysis of language is not good enough. When we introduce a term 'x', we imply that this denotes an object x that can be differentiated from other things in the world. Thus, the use of any such ordinary language already implies that the world is carved up in some way, and we must identify where to draw the right boundaries. The analysis of language and its relation to the world can therefore give us no insight into whether the world *is* pre-parcelled into distinct things or whether it only appears this way. Since, however, the brain is not like a language system, its analysis has allowed us to gain the relevant insights.

We see how a language-based approach is wrong-headed by reconsidering Blackburn's question: 'what is meant by saying that "trees" are mind-dependent, if trees are not?' The way this question is formulated implies that there is a mind-

---

[9] Note that Smith also suggests that individuation is dependent on an organism's cognitive processes (Smith, 1996, 2002), challenging standard naturalist accounts such as Dennett's.

independent *plurality* of things such as trees that affects our senses such that we have a thought of 'tree'. Implying the existence of a plurality of things again implies a certain sortal structure—namely the structure of enumerable objects. But where does this structure come from? The idealist's answer is that a subject must draw lines between the different things we find in the world, because the differentiation is already something that we do not find in the world itself. Thus, the right question to ask would be: 'what is meant by saying that the things we call "trees" are mind-dependent, if the world is not?' And the right answer to this question is: 'Any divisions in the world are divisions made by the mind rather than divisions in the world.'

While there exist absurd versions of idealism that deny the existence of anything independent from the mind, a *moderate idealism* like Kant's transcendental idealism does not claim that there is nothing independent of mind, nor that the universe only began on the day I or some other cogniser was born. Instead, it asserts that the world lacks essential structure to be conceptualised, or that it has no structure that is epistemically accessible to us. It is only thanks to an observing mind that imposes certain structures onto the world that the world gains a structure such that propositions about specific events can be formulated, allowing for reasoning and thought.

It may still be that certain structures can coherently be said to exist in the world independently. Perhaps a mathematical formalism in terms of a single energy distribution function can express all there is in a physical, eliminativist ontology. However, an essential ingredient to all our linguistic thought and to the formulation of facts we find in the ordinary world is the mind processing the data. It is the object–property structure we find in the world which we impose onto it by how our mind operates. This view challenges the realist's metaphysical endeavour because the object–predicate relationship is the crucial structure we require for *any* formulation of facts in the material world.

Given the consequences that arise for metaphysical thought, the idealist position cannot be said to be banal or empty. Once the Px-structure is said to result from our cognition, then subsequent questions about the structure of reality must also be analysed from the perspective of cognition. This includes questions regarding the puzzles within the manifest image (which will be discussed in the next chapters) as well as the clash between the manifest and scientific image. The idealist claim would be somewhat banal if there were no further consequences or explanatory benefits of this view in contrast to realism. But precisely the opposite is the case. In earlier chapters, I discussed that, according to quantum mechanics, localisation, separation, and identification of ordinary objects do not, according to some, find an equivalent in quantum physics. Let us briefly consider how our understanding of neural network processing allows us to explain how these three aspects of ordinary objects come about. This illustrates the benefits of an idealist metaphysics.

13.4 Simulating Separability

Simulations can help explain how the perception of separate objects is possible. In a toy example shown in Fig. 13.4, the data (Fig. 13.4a) consist of images that humans perceive as either ‘0’ or ‘2’. We perceive these images as such because our neurological system has already created an internal representation of these categories. However, each instance of the data consists only of a number of pixels with varying pixel values, and there is no clear separation between the zeros and twos. Nonetheless, some neural network models (Fig. 13.4b) can make clear separations of the data over the course of a learning process. This can be achieved in an unsupervised manner, without any labelling of the input images. In the case discussed here, the input image is set to be the output target image (in an autoencoder fashion). Therefore, the objective of the network is to recreate the input values as accurately as possible by building an internal representation of the data (Fig. 13.4c) and generating an image from this representation. The ‘goodness’ is determined by an objective function—a function that returns a high value when the model’s input and generated output strongly diverge. This function should be minimised by learning an adequate representation of the data.[10]

Based on such learning techniques, unsupervised image segmentation algorithms have become very popular in the computational vision community.[11] After learning an internal representation of the world, these algorithms can distinguish an object from its background. Given such a representation, the usual approach is to compute, for each pixel, the probability of being connected to the surrounding pixels. The upper section of Fig. 13.5 displays some images of objects that individuals might

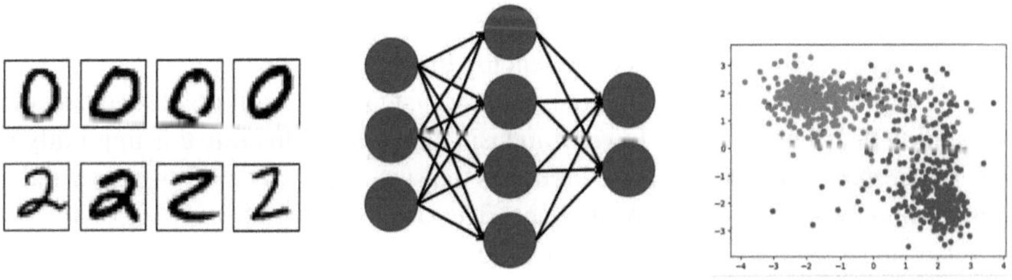

(a) Input (b) Model (c) Latent space

Fig. 13.4 The raw input data (**a**) are transformed by the model (**b**) into two separate clusters in the network’s representation or latent space (**c**). Those points that are far from the cluster centres correspond to images presented to the model at an early stage in the learning process, when the model has not yet managed to build its internal categories

[10] This is similar to how we think the brain learns about the world. See Rao and Ballard (1999) for further details.

[11] For a survey on unsupervised segmentation algorithms, see Zhang et al. (2008). See also Spratling (2013) for an image segmentation model of the cortical area V1.

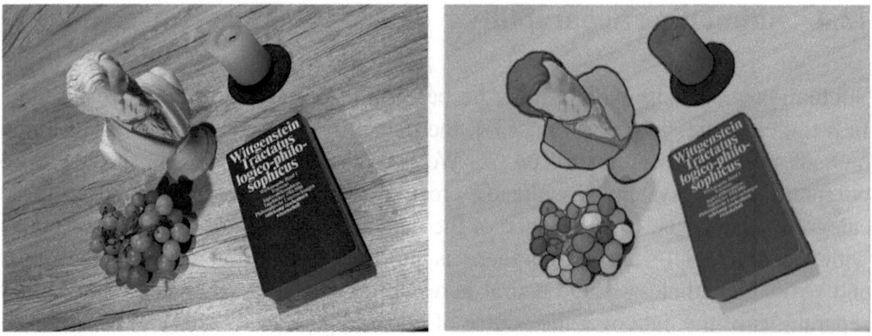

Fig. 13.5 An image segmentation algorithm uses the left image as input and returns the right image as output. Image generated with the algorithm of Kirillov et al. (2023)

encounter in their daily lives. Below, these objects are distinguished from their backgrounds algorithmically. Within the machine learning community, there exists a large number of different models for how such an algorithm can be implemented in supervised or unsupervised fashions.

We see from such simulations that we need not assume distinct, mind-independent objects in our ontology to make sense of image segmentation and separability. A system that perceives the world by segmenting images will be inclined to believe that separable ordinary objects or substances exist independently of cognition—even if this is not the case in some mind-independent, quantum-mechanical sense.

Simulations conducted in computational neuroscience research provide strong reasons to believe that there is an advantage for organisms to see objects as individuals. I will briefly discuss some results that arguably indicate how an evolutionary benefit arises from experiencing the world as we do.

By sampling the world through the sensory system, the brain maps a continuous probability distribution onto a lower-dimensional space. I illustrate this implicitly in Fig. 13.4, where each image in the input has the size of $28 \times 28 = 784$ pixels and is thus a point in a 784-dimensional space. The data, consisting of a set of images of zeros and twos, are samples from probability distributions in this very high-dimensional space. Computationally speaking, the brain has two options: it either explicitly encodes an approximation of this probability distribution or it encodes the distribution's different states. Today, it is strongly believed that the brain does not encode the full probability distribution, but only the different states expressed by the distribution's peaks. This is computationally realised by introducing a non-linearity (e.g., a threshold) that distinguishes the different peaks of the probability distribution (e.g., based on the local maximum likelihood).

By simulating point neurones, Mihai Petrovici (2016) computationally explains our experiences when we are confronted with certain optical illusions. Consider the famous case of the duck-rabbit. A subject looks at the duck-rabbit image and alternates between two different image interpretations. This situation is considered to indicate that the network activity encodes states, rather than probabilities. The brain is either in one state or another, and not in an intermediate state. If the reverse were the case, we would not see either a duck or a rabbit—rather, our brain would judge the duck-rabbit to be a superposition of both a duck and a rabbit, with different probabilities (e.g., 40% duck and 60% rabbit). There are at least two simple reasons why such a neural implementation would not be useful for survival. First, as Petrovici demonstrates, computing multiple possible states is computationally very inefficient (Petrovici, 2016, ch. 6.4), resulting in wasted resources. Second, the brain must be committed to a decision to act in response to one of the categorical alternatives indicated by the peaks in the distribution of the probability mass function. If the visual cortex computes that there is a 10% chance of a donkey, a 30% chance of a house cat, and a 60% chance of a hungry lion standing next to me, then there is little benefit in adjusting my behaviour according to the first two probabilities. Therefore, bringing the brain into a determinate state was favoured over the course of evolution.

13.5 Illustrating Communicating Systems with Simulations

Above, we have seen Blackburn's concern with idealism. A second typical worry about idealism is that if mental processes produce the essential structures of our perception, how then is communication between different people possible (see, e.g., McDaniel, 2020, ch. 3.3)? And how can different species individuate the same object? When I throw a stick, my dog runs after it because it tracks the same individual. Is this not only possible because there really is this object in the world?

Clearly, different living beings experience the same structures in the world in a way that enables communication and interaction to function. The explanation for why we experience objects in a Px-structure assumes Friston's definition of what it means to be a living being. Thus, any living being is expected to see objects in the Px-structure. If one wonders why individuation is not different for different living beings, despite having the same structure, the answer is that they live in a similar environment, where some forms of individuation are more successful than others. This does not, however, mean that the Px-structure itself is part of the mind-independent world.

For illustration purposes, the communication process can be simulated in the following manner (Fig. 13.6). One develops two agents' PP models, A and B, each of which can identify an object and generate an image from the corresponding class. This self-generated image is shared with the other agent, who ideally identifies the image correctly. This works even if the two agents have slightly altered neural network architectures—a plausible assumption given that no two cognitive systems are absolutely identical. Both agents can be trained on different sets of the image database to illustrate that communication does not require exactly the same prior

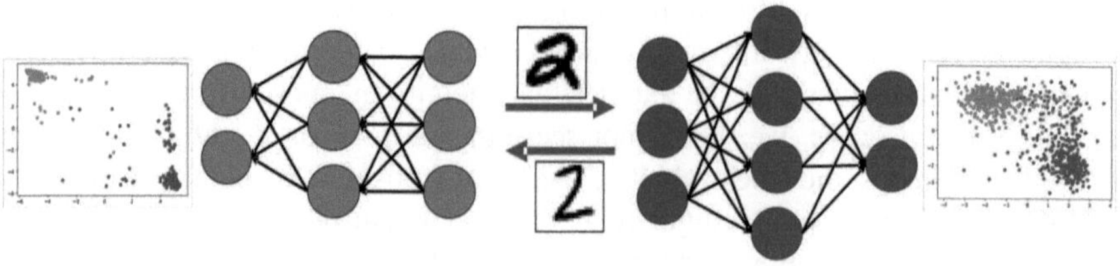

Fig. 13.6 Two trained neural networks can communicate with each other by receiving generated images from the other network and returning images of the same class

experience. When A chooses a sample point within a region of high density in its latent space, it generates an image via top-down computations over the network. This image is shown to B, who *identifies* the image by forward, upstream propagation and samples a new self-generated image from the corresponding class by top-down generative image processing. This alternation between generation and identification of images is a process that allows agent A to be informed that agent B has sampled the correct internal class representation, and vice versa. Thus, both agents possess the information that they are communicating about the same object, even if no such *thing* actually exists independently of their joint construction.

Simulations such as those presented here indicate that arguments aiming to show that we must presuppose objects in our mind-independent ontology fail. There need not be any *thing* out there in the mind-independent world that serves as a precondition for my capacity to communicate and experience the world as I do, in terms of separate, individual objects. The only crucial point is that *all communicating agents share a similar cognitive process* with regard to the underlying computations and are exposed to a similar environment. This enables the agents to understand each other if communication is modelled as described above.

It should be noted that some realists think that it 'seems plausible to suppose that there is nothing we can say that the Martians can't' (van Inwagen, 2009, p. 479), and vice versa. If 'Martian' is interpreted as any living system in any possible universe, then this statement is false. Although the Martian is a living being and thus experiences the world through a Px-structure, it may still be the case that the Martian's information-processing system is encoded very differently and will not necessarily be able to have the same categories as humans do. Thus, it will not necessarily individuate the same objects. The No-Free-Lunch theorem, which we have encountered earlier, tells us that it is even possible for a Martian never to grasp the concept of a teacup, regardless of how many cups I present it with, and that the Martian will not be able to recognise the cup as an individual, given only its perceptual system.

Summary

The perceptual process of individuating objects and distinguishing them from their backgrounds, as well as from other objects, can be explained as a consequence of

PP without making assumptions about the world's mind-independent objects. In the form of a debunking argument, this implies that it would only be by coincidence if the Px-structure were also present in the mind-independent world. The resulting idealism can defend itself from concerns about triviality and about failing to account for communicating individuals.

Chapter 14
Explaining Manifest Structures

From the problem of underdetermination, we have seen that for a given cogniser that receives sensory data, there is a *multiplicity of possibilities* regarding what the experienced qualities of ordinary objects could be and how the visual ontology is structured. Using feedback loops that add information from different parts of the brain to the information coming from the senses, the lacking structural information in the data is complemented by assumptions made by the brain about the world. What is investigated in this chapter are these additional structural aspects that our brains impose on the sensory data for the purpose of seeing objects in a way that is beneficial to our needs and, due to underdetermination, not necessarily veridical. Thus, *given* our visual sensory input, this chapter suggests that the structural aspects of our visual ontology are a product of how the brain computes sensory information. Whether the universe truly consists of particles in motion that persist over space and time, or whether we occupy a universe of cellular automata that are motionless (see, e.g., Wolfram, 2002), is a question that might at some later point be decided on the basis of new scientific evidence. While our best physical theory might someday turn out to postulate motionless cellular automata, our ordinary sensory data cannot provide us with an answer to these fundamental questions of reality. However, it is clearly the case that we experience ordinary objects as persisting over space and time, and this chapter offers an explanation of how the brain adds this structural element to the given visual sensory data, thereby shaping the intuitions we have about ordinary objects and explaining how some of our metaphysical concepts are formed.[1]

This chapter is derived in part from an article *Predicting ordinary objects into the world* by Arthur C. Schwaninger, taken from *Philosophical Psychology* ©2022 The Author. Published by Informa UK Limited, trading as Taylor & Francis Group.

[1] As we have seen, some of the qualities we associate with ordinary objects are not only underdetermined by perception but even inconsistent with physical reality (e.g., the inseparability of individuals due to quantum entanglement).

A. C. Schwaninger, *Cognitive Metaphysics*, Synthese Library 529,
https://doi.org/10.1007/978-3-032-05420-3_14

Humans experience ordinary objects, such as chairs and tables, as things that endure over space and time, that are not collocated with one another, that are composed of proper parts, and that survive the loss of some of their parts. Multiple articles in contemporary theory of perception attempt to make sense of how our visual representations generate the 'visual ontology' (Skrzypulec, 2016, p. 261) of our experiences (e.g., Green, 2019, 2021; O'Callaghan, 2016; Skrzypulec, 2016, 2018). By turning to findings in the cognitive sciences, they explain why it is that we see certain parts of our visual field as composing single objects with the properties we experience them to have. These approaches are reminiscent of Gestalt psychology, which seeks to identify the rules that the mind employs to generate perceptual units.

The goal of this chapter is, first, to explain how the category of *robust predictive processes* (RPPs) allows us to unify classical concepts used to characterise the objects in our visual ontology, such as compositionality and persistence, under a single umbrella term. Second, using this new concept, the chapter explains why it is that the sensory data are interpreted by the brain as having the structure that we experience. It shows how the category of RPPs unifies different aspects of ordinary objects in our visual ontology and outlines the advantages of employing RPPs in our theorising.

14.1 Predicting Essential Structures into the World

I show how certain structural properties of our visual ontology cannot be derived from the visual sensory data alone but are, to some extent, the result of our predictive cognitive processes. While there exist numerous ways in which our visual data could be interpreted, it will be shown that the discussed structural qualities of ordinary objects can be derived from the category of an RPP, given our ordinary sensory input. The category of RPPs is therefore argued to unify certain concepts within our visual ontology.

The postulated principle governing the visual ontology of a system A, which has a cognitive system that operates by the principles of PP, is the following:

Principle of Predicted Structure A's visual ontology is structured in a way that robustly maximises the probability of a minimal average prediction error for future sensory input.[2]

At this point, the notion of robustness requires further explanation. According to PP, higher cortical regions in the brain actively generate percepts on the basis of Bayesian probability estimates, trying to *predict* the sensory data in a *robust* way. As the cognitive system does not rely on innate feature detectors, it is highly flexible

[2] In a noisy world, robustness is in the generic case already implicitly entailed in minimising the average prediction error. However, for the purpose of emphasis and generality, it is made explicit.

in adapting to a continuously changing environment and can use its resources effectively. We have seen that this is because the information-theoretic complexity of the world is far greater than the complexity of the brain (despite the brain being the most complex isolated *biological* system we know), and that the brain must find a model representation of the world that compresses all relevant information.

If it were, theoretically, not the case that the world surrounding us had a much greater complexity than our brains—if, in other words, there were no constant unforeseen changes in the environment—then the PP strategy of cognising the world would be far from optimal. For instance, a probabilistic approach to cognition would make little sense in a universe with only a handful of possible states and a simple rule determining the upcoming state, if the cogniser had more than enough resources to process these states.[3] Thus, there is an implicit ontological assumption about the world surrounding us towards which PP is adjusted, namely, that the world is much more complex than the cognising system itself and is constantly changing, making innate feature detectors pointless. In such a highly complex world, the cognising system is by definition limited in its capacity to grasp the full complexity of its environment. As a consequence, much of what is going on cannot be comprehended by the cogniser and is perceived as *noise*. Living in a complex world therefore necessitates cognition that can deal with noise if it is to be successful.

This is where the significance of *robustness* comes into play. In a complex world, the cogniser is surrounded by noise and must therefore maintain an internal model of the world that is largely unaffected by it in order to minimise the average prediction error. For a predictive cognitive process to be qualified as making *good* predictions, the process must not only exhibit high accuracy and precision when applied to generic sensory data; it must also make successful predictions when environmental changes occur or when the sensory data are convoluted with noise due to the cogniser's biological constitution. Such predictive processes are then said to be *robust*.[4]

For instance, predictions of neural activity should not differ drastically whenever shapes in the perceptual data are rotated or when luminance varies. In the case of visual perception, the perceptual capacity of being robust manifests in *perceptual constancies*. *Shape constancy*, for instance, is the capacity to see a shape under a variety of different perspectival conditions. *Location constancy* is the capacity to detect an object at varying distances (Burge, 2014). Perceptual constancies are to be understood as a consequence of minimising prediction errors in a robust way. In the PP framework, this robustness is ensured by the hierarchical structure of the network, in which higher cortical regions predict the activity of lower regions.

---

[3] For simplicity, the cogniser is, in this example, implicitly assumed not to be part of the world.

[4] In the machine learning literature, the lack of robustness is characterised by and related to the concept of 'overfitting'. A learning algorithm that attempts to take too many possible instances of a hypothesis class into account does not generalise the data well but instead overfits it, resulting in a lack of robustness to new data.

We have, in the previous chapter, already encountered the *Principle of Individuation*. This principle follows directly from the *Principle of Predicted Structure*. In what follows, further principles will be presented that can also be derived from the *Principle of Predicted Structure* and concern specific aspects of our visual ontology.

Object Persistence

When it comes to the phenomenon of object persistence, it has not gone unnoticed that the engagements between philosophers and scientists are related:

> What makes an object the same persisting individual over time? Philosophers and psychologists have both grappled with this question, but from different perspectives – philosophers conceptually analysing the criteria for object persistence, and psychologists exploring the mental mechanisms that lead us to experience the world in terms of persisting objects. It is striking that the same themes populate explorations of persistence in these two very different fields – for example, the roles of spatiotemporal continuity, persistence through property change, and cohesion violations. Such similarities may reflect an underlying connection, in that psychological mechanisms of object persistence . . . may serve to underlie the intuitions about persistence that fuel metaphysical theories. This would be a way for cognitive science to join these two disparate fields, helping to explain the possible origins and reliability of some metaphysical intuitions, and perhaps leading to philosophical progress. (Scholl, 2007, p. 563)

It is clear from introspection that, within our visual ontology, objects persist over some time and space, and we expect them not to disappear from one moment to the next without any reason.[5] This means that the corresponding parts in the visual field do not become a new object from one moment to the next, but persist as the same object. In Fig. 14.1, an example of the visual image over a short amount of time is presented. Due to the underdetermination of perception, there is no empirical fact

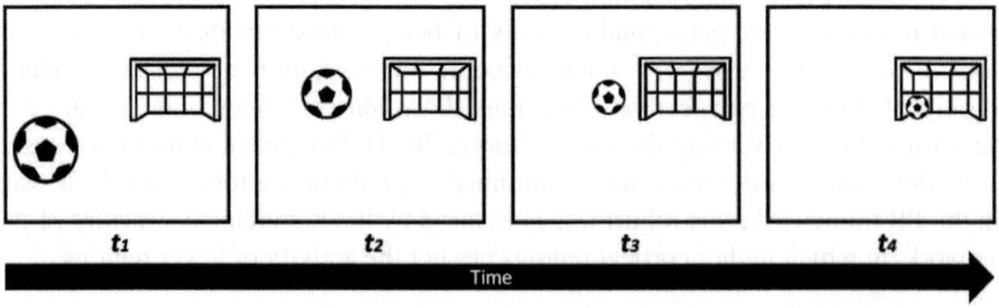

Fig. 14.1 Objects are experienced as persisting through time and space even though the sensory data do not necessitate the connections between the objects at different points in time

[5] Studies show that infants already expect objects to move along continuous trajectories and are surprised if they do not do so and disappear (Spelke et al., 1995).

of the matter that necessitates the connection between the ball in one image and the ball in the next, since our senses receive sequences of sensory data that do not determine whether the objects at one moment in time are the same as those in the next. Yet, we clearly have the impression that the ball remains the same ball—the same object.

The *Principle of Individuation* explains how an object x comes about in our visual ontology. Assuming that we have at least one such object, then, according to the postulated *Principle of Predicted Structure*, the following principle must hold for any predicting cognitive system A, since persistence is a structural property:

Principle of Object Persistence An object x that is located at the spatial-temporal position $(\mathbf{r}_1, t_1)$ continues to be perceived by A as the same object at $(\mathbf{r}_2, t_2)$, if this robustly maximises the probability of minimising A's average prediction error.

The important question is whether the *Principle of Object Persistence* correctly captures our intuitions concerning object persistence, namely that ordinary objects generally persist through space and time, given the visual images we receive on our retina. Thus, with regard to Fig. 14.1, the question is whether the *Principle of Object Persistence* provides reasons as to why we see the ball at one point in time as being the same ball at the next point in time.

Giving an affirmative answer to this question is almost trivial. When one considers the opposite scenario of a cogniser who experiences ordinary objects as constantly coming into and going out of existence, the cogniser would be unable to make any useful predictions about upcoming sensory inputs. In the case of the ball in Fig. 14.1, after times t_1 and t_2, the best prediction that minimises prediction error at t_3 is *not* that the ball at t_1 and t_2 suddenly disappears and a new ball appears at t_3. Instead, a trajectory of what the cognitive system identifies as a single object is computed, and predictions are made about where the ball will land at t_3. Therefore, according to the interpretation of PP presented here, ordinary objects persist within our visual ontology whenever the visual data allow the brain to make predictions.

Parthood and Compositionality

The different aspects concerning parthood and compositionality extend the case of object persistence. The central question that we are concerned with here, regarding our visual ontology, is how we account for the fact that some parts of our visual image compose a unified object, while others do not. For instance, how can it be that my arm composes an object with the rest of my body, but my nose and the Eiffel Tower do not compose?

Within the metaphysics literature, we have seen that van Inwagen (1990) asks under what conditions two objects, x and y, jointly compose an object, z, and calls this the special composition question. Clearly, as a realist metaphysician, van Inwagen seeks an account of compositionality for a 'mind-independent, discourse-independent world' (Horgan, 1993, p. 695). Our interpretation of the question here is a different one: it concerns how, in the spirit of Cognitive Metaphysics, the brain processes the visual ontology of our experiences—not what metaphysicians have in mind when talking about compositionality in a mind-independent world. As we have seen, compositionality is itself a manifest concept, and thus it makes sense to understand metaphysical work also in relationship to our neurological architecture and to engage with metaphysical puzzles from this perspective. Besides the special composition question, we also encountered problems like Unger's (1980) problem of the many, for which we have seen partial solutions from an information-theoretic perspective—partial because they result in undesirable consequences, but at least imply a second-order property of maximality that can be attributed to a property F, such as *being a cloud*.

What I intend to illustrate now is that the *Principle of Predicted Structure* implies maximality when the cogniser is exposed to our ordinary sensory data. It is for this reason that our brain avoids Unger's problem and that we have the experience of seeing only a small number of clouds rather than millions of them when looking into the sky.

Notice that the *Principle of Predicted Structure* implies the following corollary about the structure of our visual ontology:

Principle of Compositionality The visual objects x and y compose to z if z is expected to be, on average, better predictable than x and y independently.

Thus, categories are selected and imposed onto the data such that regions in the visual image turn into individuated objects we perceive as single, unified objects. The important thing to notice about this principle is that it entails Sider's maximality property for the objects of our visual ontology when we assume that sensory data is noisy and comes from a complex, changing environment, as presupposed by PP.

To see this, consider Fig. 14.2. A black square is depicted, which is moving around in front of a cat, thereby only allowing the cogniser to garner a small number of visual cues about there being a cat in the data. Such a lack of information and noise in the data is commonplace in ordinary, everyday perception. Every time we

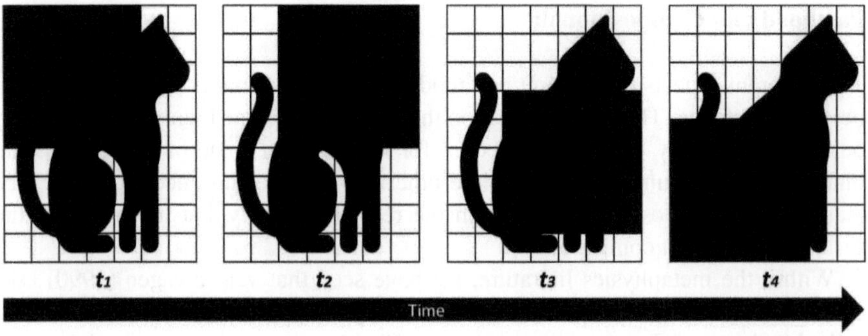

Fig. 14.2 The cognising system must select the category that is expected to be the most robust against noise over time in order to minimise prediction error

walk around or move our eyes, there is a 'dramatic change of sensation' (Frith, 2007, p. 201), and yet the world appears to be stable.

Let us first assume a cogniser with a cognition process that assigns categories merely by making a similarity comparison of the visual image with some static features or prototypical image in memory, as in the above-discussed naïve approach to cognition. In this case, at time t_2, the object in the image could just as well be interpreted to be a dog, because at that moment in time the relevant information that makes a distinction between dogs and cats (the head) is unavailable, and the body of the animal might just as well be a dog's body. (For simplicity, I am assuming that the sole relevant visual characteristic distinguishing the animals is given by the shape of the head.) This type of cognising system will have a visual experience of seeing a cat, then a dog, and then a cat again. Or, in general, the system will have an unstable visual experience of a world in which categories, as well as the boundaries of objects, alter from one moment to the next.

In contrast, a cognising system that has experiences by minimising prediction error will experience a visual ontology of stable, non-abruptly altering objects and object kinds. It will impose the category onto the data that is the most robust towards possible future noise in the sensory data, because this makes the data predictable. The cogniser's category would, for instance, not change at time t_2 to 'dog', because this would not be a robust prediction over time. Any categoric change leads to a high prediction error, since a category is needed in the first place to make a prediction about an upcoming sensory event.

This so far extends what we have already seen when looking at object persistence, but now the important point is that our findings generalise to the parts of the object. To see this, let us again assume a cogniser with a cognition process that assigns categories by making a similarity comparison of the visual image with some prototype in memory. This cogniser is confronted with the sequence of images in Fig. 14.2. As we have seen, the object in the image could, at time t_2, just as well be interpreted to be a dog. But furthermore, some small parts, for example, the tip of the cat's claw, might not be well in line with the prototype case of a cat and would thus be treated as a separate object. Or the whole region of the cat minus some small part A could look equally alike to the prototype as the cat minus some small part B. Therefore, the cogniser might interpret there to be two cats at hand, or randomly select one possible instance over the other.

Furthermore, imagine that the cognising system were, for instance, not to select the category of the cat as a whole at time t_1, but instead were to impose the categories of cat-head, cat-body, and cat-tail separately. In this case, there would be a great amount of prediction error at time t_2, when the square overlaps the head, as the activated category of the cat-head would have to be turned off completely (including any further encoding for the spatial connection between the cat's head and all its other bodily parts). At t_3, the representation of the head would have to be reactivated, generating further prediction error due to the sudden reappearance of the object. If, on the other hand, the cognising system were to impose the category of a *whole* cat at t_1 and maintain this categorical assumption for as long as it is a reasonable option, then no great prediction error would be expected at t_2 or t_3, making this the category of choice. A selected category that is not robust against such noise is not reliable in minimising prediction error, because in daily perception we often only receive a glimpse of an object.

Contrary to a cogniser that selects categories by naïve similarity comparisons, a cogniser operating on the basis of making robust predictions will always see the maximal number of jointly predictable parts in the visual field to compose as one object, because this increases the chances of making good predictions from prior experiences. Such a cogniser will look into the sky and see *one* cloud, rather than each part of a cloud forming an additional cloud as considered in the problem of the many. In contrast, a naïve categorisation by similarity analysis leads to a multitude of possible categories, where different parts of the visual image might match the category well and the boundary of the object would not be well determined.

To be sure, a cogniser's perceptual system that categorises for the purpose of making robust predictions will, on the one hand, see objects as composing wholes (e.g., one cloud rather than millions of clouds), but will nonetheless not see objects composing arbitrarily. My nose and the Eiffel Tower do not compose, because the mental category needed to individuate such an object does not lead to robust predictions. There is no connection between my nose and the Eiffel Tower that would result in any better prediction error minimisation, were they to compose to a single object.

Collocation

From the *Principle of Individuation*, the following must hold: one and the same region in the visual field is simultaneously individuated as two different objects x and y if this robustly maximises the probability of minimising A's average prediction error. Clearly, we humans *never* experience objects to be precisely collocated. This becomes most apparent when we think of the famous duck-rabbit illusion (Jastrow, 1899), in which we are unable to see both a duck and a rabbit at the same time in the same location.

The fact that visual objects are never seen to precisely coincide does not follow trivially from the above-mentioned principle for our sensory input. As we have seen, for a PP approach to operate successfully, the ontological assumption must be made that the environment is complex and in flux. It is sometimes suggested that additional ontological factors determine the fact that we do not see ordinary objects to coincide. For instance, in the context of explaining *binocular rivalry*,[6] which arguably also applies to the duck-rabbit illusion, Hohwy et al. (2008) suggest that our generative top-down model has learned that 'only one object can exist in the same place at the same time' (p. 691). According to them, our experiences of the world shape our brain's prior expectations to the extent that, by Bayesian inference,

[6] This is the phenomenon whereby, when the left and the right eye are presented with different images, the subject alternately sees only one image at a time rather than both.

the brain computes a much lower probability for the image to be both a duck *and* a rabbit rather than a duck *or* a rabbit.[7]

The problem with Jakob Hohwy et al.'s analysis is that they only maximise probabilities for a given moment in time that is case-specific rather than general, and thereby fail to account for general robustness over time. In accordance with PP, the cognising system A has no innate categories of perception, but is rather constantly in the process of constructing these categories to account for the many small changes in the environment. As we have seen in the case of compositionality and persistence, if A were to alter its hypothesis every time some other hypothesis seems more probable, it would incur an extremely high prediction error in a changing and noisy environment. It would look at a cloud and suddenly see a huge horse appear and then, in the next moment, a huge unicorn, due to constant formation changes in the cloud. This is precisely what any system based on PP seeks to avoid. And given that categories are not innate or predetermined, any indecisive decision in which the cognising system begins to perceive a multiplicity of possible objects for the same visual input generates complete instability in the system.

Minimising prediction error in the long run comes at the cost of not being overly considerate about alternative possibilities in the moment. For this reason, the brain must occasionally disregard hypotheses even if they have a very high probability of being the optimal choice in the moment, or are comparably good to another. The first consequence of this is that the following principle about collocation must hold:

Principle of Object Collocation One and the same region in the visual image is never simultaneously individuated as two different objects.

The second consequence is that having a discomforting visual experience like the duck-rabbit illusion is 'something any organism with a predictive engine will to an extent experience' (Kaaronen, 2018, p. 11), and surely not a socio-culturally acquired drive (as suggested by Cooper, 2007, pp. 87–89). The illusion amounts to a general phenomenon that is universal for all living cognisers in the universe.

It might seem that the fact we do not see ordinary objects coinciding with each other results from the physical fact that the subatomic particles constituting ordinary objects also do not coincide. Therefore, one might try to argue that the fact that ordinary objects do not coincide in our experience is the result of how the world really is, rather than being a consequence of how our cognition works. However, this again misses the point that the empirical data are underdetermined, and there exists a multiplicity of possibilities for interpreting the world. For instance, all empirical data in physics are also compatible with the assumption that each particle in the world precisely coincides with a massless 'dummy' particle that does not causally interact with anything. The data might also be compatible with each particle in the world having half its mass, but always being collocated with another particle of equal mass. In any case, our cognition decides for one of many possibilities, and

---

[7] Notice that I do not equate 'seeing a duck and a rabbit at the same time' with 'seeing a duckrabbit', as this might be confused to refer to a single object, a duckrabbit, rather than two coinciding objects.

the fact that we experience ordinary objects to be necessarily non-collocated in any complex and changing world is a fact that can be derived from how cognition works, rather than from the empirical data.

14.2 Reformulating Questions About Ordinary Objects

The analysis shows that any system that receives our ordinary visual input and experiences the world on the basis of RPPs will experience the world to be structured in the following way: it consists of individuals that tend to persist over space and time, that are not collocated with each other, and for which maximality is guaranteed despite compositionality not occurring arbitrarily. Problems such as the Problem of the Many are, within our visual ontology, an immediate consequence of a simplistic view of cognition that relies on the intuitive assumption that categories must be applied to objects based on the similarity in the arrangement of their parts. But this is not how the brain subconsciously categorises objects.

We notice that the notion of an RPP allows us to unite several aspects of what it means to be an ordinary object under one umbrella term. This provides a more unified understanding of the object concept, which is to be preferred over other accounts of our visual ontology that depend on numerous independent principles (Green, 2019, 2021; Skrzypulec, 2016). Questions concerning our visual ontology that are ordinarily formulated in terms of standard metaphysical concepts can now be reformulated in terms of RPPs by taking a subjectivist position and moving from a mind-independent metaphysics to a metaphysics that relates to epistemology and perception. For instance, instead of asking 'Do x and y compose?', we can ask 'Are x and y jointly better predictable on average than independently?' Or instead of raising the question 'Does x persist through space and time?', we might ask 'Does the individuation of x also lead to good average predictions if x is located at some other location in space and time?' Or 'Does x survive the loss of its part y?' can be replaced with 'Does x without y still lead to small average prediction errors?' While the traditional questions have no straightforward answers and seem not to be directly related, or at least hard to connect, the new formulation relies on the unifying category of RPPs and offers a straightforward answer. This is not to say that no implicit ontological assumptions were presupposed in our analysis, since we presupposed the visual input we already ordinarily have. However, due to underdetermination, certain cognitive import is required to give structure to our visual ontology. The concept of RPPs unifies the missing structural aspects and also allows us to explain in a unified manner how our intuitions about our visual ontology come about, given our ordinary sensory input.

In the context of Cognitive Metaphysics, the concept serves as a fundamental *metaphysical* category that gives structure to the reality we experience, or describe indirectly, in most of our scientific descriptions. RPPs structure the world not instantaneously but form it over the course of a process, implying that the structures of reality must be understood in processual terms. This means that while we always see the same type of structure, how exactly a given visual input is structured changes from one moment to the next, depending on how the neural processes deal with the data. While objects are always individuals persisting over space and time, it is not necessary that a given image is always interpreted to contain the same objects. Thus, while RPPs are the fundamental category giving shape to our experiences, the objects themselves are subject to cognitive variability.

Summary
Ordinary objects are experienced to endure over space and time, to not be collocated with each other, to be composed of proper parts, and to survive the loss of some of their parts. These qualities are, on the one hand, difficult to reconcile for theorists of perception, and on the other hand, pose a variety of problems when considered in isolation. Relying on the theoretical framework of PP, this chapter argues that we can use the category of an RPP to conceptualise qualities such as persistence and compositionality in a unified manner. Traditional problems concerning the structural properties of ordinary objects, such as the question of when two objects compose, can then be reformulated using this new category and answered in a straightforward manner. What follows is an idealistic process ontology, where RPPs shape the objects surrounding us.

Chapter 15
Naïve Classification and the Nature of Vagueness

For the idealist, the way the cognitive system categorises our experiences is crucial not only to our perception but also to the formation of the world. For the realist, however, categorising is merely labelling the objects that are *out there*, given to us in their mind-independent structure. A highly attractive interpretation of how this labelling might occur is what we have called the *naïve view on classification* (see p. 158) by which the brain labels objects according to their similarity with other objects or with some prototypical object instance. Hume is one of the philosophers who promoted the naïve view and contributed to its wide spread acceptance. In his seminal work, *A Treatise of Human Nature*, he explores the origins and nature of our ideas and abstract thinking, reflecting on how the mind forms general ideas from more basic ones. He argues that minds have a natural tendency to notice similarities among distinct objects which he explains as follows:

> When we have found a resemblance among several objects, that often occur to us, we apply the same name to all of them .... After we have acquir'd a custom of this kind, the hearing of that name revives the idea of one of these objects, and makes the imagination conceive it with all its particular circumstances and proportions. (Hume, 1739, pp. 18–19)

While this formulation of the naïve view on classification appears compelling, it immediately results in a paradox known as the *sorites paradox*. The paradox's first formulations date back to the Megarian philosopher Eubulides from the fourth century BC and are best understood through a concrete example involving a sequence of slightly changing objects:

> A man with no hair on his head is bald. Add one hair to his head and he is still bald. Add another hair to the man's head and he will still be bald. Continuing to add one hair at a time, it seems that each individual hair cannot transform him from being bald to not bald. The paradoxical conclusion is that a man with a million strands of hair is bald.

Even though all men at any location in the sequence resemble each other, the first and the last man are classified differently with regard to the predicate of baldness. Hume's belief that objects are named based on their resemblance leads to the absurd conclusion that all objects could be classified under a single name. This is because

A. C. Schwaninger, *Cognitive Metaphysics*, Synthese Library 529,
https://doi.org/10.1007/978-3-032-05420-3_15

there will always be a sequence of objects where each one closely resembles the previous, even if the first and last objects in the sequence are vastly different. In the case of the bald man, the paradox can be formulated as a chain of statements about a sequence of objects as follows:

A man with 0 strands of hair is bald
If a man with 0 strands is bald, then also a man with 1 strand of hair
If a man with 1 strand is bald, then also a man with 2 strands of hair
If a man with 2 strands is bald, then also a man with 3 strands of hair
...
If a man with 999 999 strands is bald then also a man with a million strands of hair

---

Therefore, a man with a million strands of hair is bald

This sequence of objects, each of which is a slight alteration of the previous one, is known as the *sorites sequence*. In this case, it is the set of objects starting with an indisputably clear case of a bald man and ending with an indisputably clear case of a non-bald man (Fig. 15.1).

The argument's reasoning can be succinctly captured in two premises and a conclusion as follows:

P_1: A man with one strand of hair is bald
P_2: Add one strand of hair to a bald man and he will still be bald

---

C: A man with a million strands of hair is bald

Or, in logical terms:

P_1: Fx_0
P_2: For all x, if Fx_i then Fx_{i+1}

C: $Fx_{1\,000\,000}$

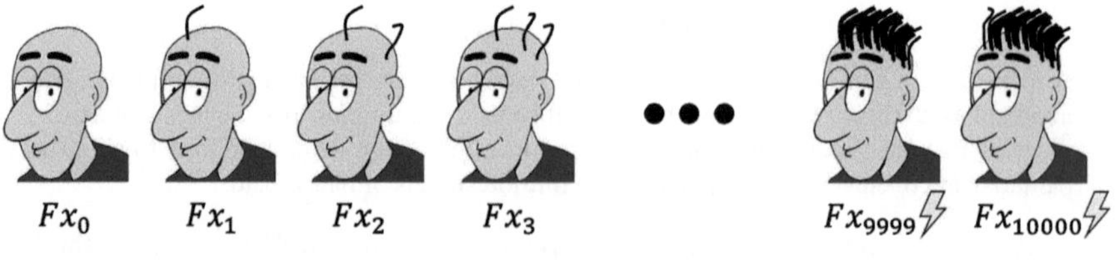

Fig. 15.1 A sorites sequence running from a typical case of a *bald man* (indicated as predicate F) to an object that does not appear to be bald at all

Although logicians seem particularly obsessed with their hair loss—given how often this example appears in the literature—the sorites paradox is not merely a barber shop dilemma. It extends, fascinatingly, to virtually any predicate referring to a material object within the manifest image. Consider formulating the paradox for the Colosseum in Rome:

P_1: The Colosseum consists of a few million bricks

P_2: Remove one brick from the Colosseum and it still remains the Colosseum

---

C: One brick constitutes the Colosseum

More generally, it appears that for any material object or kind of object of reasonable size, it is possible to sequentially remove tiny cubes of matter—say 1mm^3 in volume—until little remains of the original object. Since any such sequence can be said to constitute a sorites sequence all ordinary objects allow us to construct an argument leading to a paradoxical conclusion: that one grain of sand constitutes a heap, one small piece of wood a table, or a pinch of dust a planet in our solar system.

There are, however, some objects for which no sorites sequence can be constructed. These include abstract objects such as numbers, as it is impossible to add or subtract anything from a number like '3' without changing it into a number of unequal value. They may also include material objects such as fundamental particles. By our current understanding of physics and physical taxonomy, it is, for instance, impossible to subtract or add anything to a Higgs boson and still have a Higgs boson. To the best of my knowledge, it is also not possible to construct a sorites sequence in which parts of a helium atom are gradually removed, without reaching a clear point in the chain where a mistake obviously occurs. Once again, the clash between the manifest and scientific image becomes apparent: the phenomenon described by the sorites paradox is universal amongst the objects of ordinary experience, but does not extend to those in fundamental physics.

Underlying the sorites paradox is the concept of *vagueness*. Arguably, vagueness is essentially characterised by two symptoms (Greenough, 2017). First, there appear to be *borderline* cases, x_i, for which we do not know whether Fx_i or $\neg Fx_i$. Second, the extension of the predicate appears to lack sharp boundaries or appears to be blurred, such that a small change to x_i, like the removal of one strand of hair, does not appear to alter the truth-value of the statement Fx_i. This second symptom is often captured by the *tolerance principle*: $\forall x(Fx_i \rightarrow Fx_i')$, where $x_i' \equiv x_{i+1}$ is adjacent to x_i in a sorites series. Again, F stands for a vague predicate such as 'bald' and x_i denotes the i-th object in a sorites sequence—in our examples, a man with i strands of hair or a Colosseum with i bricks removed.

The absence of vagueness at the level of fundamental physics, contrasted with its omnipresence in the manifest image, leads the eliminativist to argue that vagueness undermines our conventional ontological beliefs. For the eliminativist, the correct response to the paradox is to reject the first premise, P_1, in any sorites argument, asserting that neither the Colosseum, nor bald men, nor any other ordinary object truly exists. According to them, statements containing vague predicates are, strictly speaking, false—no matter what (Unger, 1979; Wheeler, 1979).

While the eliminativist's reply to the sorites paradox is problematic for reasons given in the first part of this book, which shows the complications we encounter when we simply dismiss the existence of ordinary objects, their response to the problem is also highly unsatisfying. Even if it were the case that ordinary objects did not exist in any linguistic framework, the paradox would continue to persist for the objects of our thoughts and would still require an answer as to what is going wrong with these objects. So how, then, do we solve the paradox? Where does vagueness come from, and what makes it so unique to the manifest image? The rest of this chapter is devoted to exploring these questions.

15.1 Earlier Attempts in Solving the Paradox

Moving along a sorites sequence, one could imagine that there is a precise transition point at which the bald man turns into being not bald. This point is called the *cut-off point*. It is the exact location between two objects along a sorites sequence at which the predicate F switches to $\neg F$. Most people would probably not expect that such a point even exists. It seems as if there is no way in which one hair could truly make the difference between bald and not bald. However implausible this might seem, a variety of philosophical theories do in fact postulate such a point.

For instance, according to *epistemicism*, most notably defended by Timothy Williamson (2002), the principle of bivalence must hold for vague predicates, and for this to be the case, there must exist a sharp cut-off in a sorites series. For Williamson, it is solely due to our ignorance and epistemic limitations that we are unable to determine where this line lies, even though it exists. The sorites is, therefore, a consequence of our epistemic limitation, and a hair that distinguishes between baldness and non-baldness actually exists. While this theory is appealing from a logical point of view, as it conveniently allows classical logic to continue to operate, it is simply not convincing because it is unclear what it is that makes a man with 432 strands of hair bald and a man with 433 strands not bald. What in the world could determine this? While Williamson thinks that it is the language community as a whole that draws these lines by their language usage and precise criteria of application, it remains unclear how the language community has the capacity for such a precise decision. Furthermore, if the cut-off point is inherently unknowable and inaccessible to anyone, it challenges the meaningfulness of asserting its existence.

Due to such limitations, many philosophers have attempted to formulate a coherent theory that introduces a penumbra region on the sorites sequence with a special truth status. For example, one might introduce a middle truth value, allowing statements to be not only true or false but also indeterminate (Körner, 1960; Tye, 1994). But this approach merely shifts the problem, creating a new question of determining the point at which a truly bald man becomes an indeterminately bald man, thereby allowing the possibility of constructing a sorites sequence between true and indeterminate. As an alternative to such a logical system with three values, one could also try to say that there are an infinite number of truth values a proposition can have, represented by a number between 0 and 1. Then there would not be an instantaneous but rather a *continuous* change in truth values when moving along a sorites sequence (Machina, 1972, 1976). But when and how exactly truth values are supposed to change, and what it really means for a man to be bald with a truth value of 0.73494, is difficult to say.

Perhaps the most serious objection to an infinite-valued logical system is that there again continues to exist a precise cut-off point between sentences taking the truth value 1 and those taking a value less than 1 (Wright, 1976; Sanford, 1976; Sainsbury, 1990).

It appears that even theories with a penumbra region of special truth values do not avoid the implausible assumption that, along the sorites sequence, there must eventually be a cut-off point where there is a sudden transition from determinate to non-determinate cases. But if this were the case, why did we accept the premises of the sorites paradox in the first place? For some reason, the cut-off point must have been hidden from us. But why does this happen, and how is this possible?

15.2 Vagueness as a Psychological Issue

While many philosophers diagnose the sorites paradox as a puzzle arising from defective linguistic conventions, this book traces vagueness and our intuitions about the induction step of the sorites paradox back to our cognitive mechanisms. As I show first, there is good reason to believe 'that the root explanation [of the sorites] lies in ... deep-seated cognitive traits' (Goldman, 1989, p. 138) and that an 'appropriate explanation could be purely psychological: some deep psychological mechanism produces the illusion of cut-off freedom' (Priest, 2003, p. 15).

The challenge towards an adequate solution to the sorites paradox is at least twofold. First, there is the *logical question*, which asks what logical framework adequately analyses the situation of the paradox without running into intractable problems. Second, according to Graff Fara, a further key challenge facing any solution to the sorites that denies the major premise of the paradox, *P2*, is what she calls the *psychological question*:

> If the [major premise] is not true, why were we so inclined to accept it in the first place? In other words, what is it about vague predicates that make them seem tolerant and hence boundaryless to us? (Graff Fara, 2000, p. 50)

The epistemicist, who believes that there is a precise cut-off in every sorites sequence (Williamson, 1994), would argue that our intuitions are formed by not knowing where this cut-off is. As competent speakers, we expect that we would know where the cut-off position is, if there were one. Hence, as we do not know where the cut-off is, we are falsely inclined to accept that there is no such cut-off, in accordance with the sorites' second premise.

A different way to address the *psychological question* comes from Diana Raffman's (1994) early approach to the sorites, which relies on a folk-psychological explanation of what happens in the mind of an individual. Given how our cognitive system works, Raffman claims that the mental state of the individual changes when inspecting different pairs of instances along the sorites sequence, resulting in changes in psychological context. In contrast to other contextualist accounts that primarily rely on communal judgement and conversational standards, Raffman's account has been acknowledged to be more basic because it requires underlying cognitive factors to explain how conversational scores are adjusted mentally (Shapiro, 2006, pp. 26–27). Raffman provides an answer to the *psychological question* by arguing, roughly, that wherever we look along a sorites sequence, adjacent pairs will be of the same category, and we are, thus, psychologically inclined to generalise these observations in the form of P_2. She distinguishes between two different tolerance principles that cognisers seem to confuse. The first of these principles is what we can refer to as the *strong tolerance principle (STP)*:[1]

STP If a sensory datum x_i looks like an F, then the adjacent datum x_{i+1} along a sorites sequence also looks like an F.

This is essentially Hume's account of categorisation, applied to sorites sequences. Together with some initial assumptions, such as x_1 looks like an F, STP will run into the contradiction of the sorites. According to Raffman, STP ought to be taken as false, whereas the *weak tolerance principle* (WTP), with which STP is ordinarily confused, is true and does not cause a contradiction:

WTP If a sensory datum x_i looks like an F, then the adjacent datum x_{i+1} along a sorites sequence also looks like an F, insofar as x_i and x_{i+1} are judged pairwise.

WTP is Raffman's version of the kind of *similarity constraint* that we find in almost all contextualist approaches and to which we will return later. In accordance with Wright (1987, p. 276), there is, for Raffman, little difference to be made between an object *being* an F and a subject's judgement of an object *looking like* an F, when we are interested in making an assessment of objects along a sorites sequence. One might be concerned that categories and their extensions are determined by the speaker-community and must, therefore, not be dependent on an individual's perceptual judgement. However, there are two points to be made here. First, even if a concept is defined by a speaker-community, it must, nevertheless, be anchored by some individuals' brain states—these individuals are then said to be *competent speakers*. Therefore, STP and WTP still hold in the extensional formulation for these individuals, given some normality constraint on the environment.[2] Second, one can argue along the lines of Stewart Shapiro (2003),

[1] The notions of a 'weak' and 'strong' principle were taken from Åkerman (2009).

[2] For example, if lighting conditions are inadequate, then a competent speaker will fail to make a competent judgement.

who claims that judgements about the extension of a predicate are determined by its meaning only with respect to non-borderline cases, and in other cases by non-linguistic factors of a psychological kind. When investigating the limits of the extension of a term along a sorites series, the socially established meaning of the term fails to draw clear-cut specifications. Instead, non-verbal factors pertaining to the individual determine the extension of non-determinate cases. In other words, there are simply no *linguistic* reasons for where and how the jump from one category to the next along a forced march occurs, but rather a decision is made in virtue of the individual's 'brute mechanical' (Shapiro, 2003, p. 71; Shapiro, 2006, p. 29; Raffman, 1994, p. 65) mental processes.

15.3 Problems with Existing Psychological Contextualism

Raffman's views rely on certain empirical assumptions she makes. According to her, WTP holds true in virtue of two cognitive processes that are constantly at work, which together are referred to as *homunculi*. One of these processes is responsible for categorising sensory data (*categoriser*), whereas the other judges whether two adjacent patches are treated the same or differently (*discriminator*). The discriminator ensures a kind of categoric robustness, preventing a categoric shift for as long as possible when moving along a sorites sequence. Raffman explains this as follows:

> As long as you judge the patches singly, your categorizer has free rein: he may, for example, categorize [x_{26}] as looking red, [x_{27}] orange, [x_{28}] red, and [x_{29}] and everything thereafter orange. When adjacent patches are judged pairwise, however, your discriminator takes over. Finding a pair to be marginally different, he will constrain his colleague to categorize them identically: "Categorize them as you like," he'll say, "but categorize them together." Thus when you judge [x_{26}] and [x_{27}] pairwise, you will place them both in the same category, no matter how they look when you judge them singly. Where [x_{26}] and [x_{27}] together look orange after the shift, [x_{26}] will evidently have shifted its category membership. If on the other hand you make all of your judgements along the series pairwise [($x_1/x_2, x_2/x_3, x_3/x_4$, etc.)], the shift from 'looks red' to 'looks orange' will occur between pairs—say, between the [x_{25}/x_{26}] pair and the [x_{26}/x_{27}] pair. In that case the shared patch (again, [x_{26}]) will shift its category membership from one pairing to the next ... Whereas [discriminatory judgements] pertain to comparisons of two presented patches, [categorial judgements] presumably pertain to "comparisons" of a presented patch with some sort of standard or prototype in memory. (1994, pp. 47–48)

Not only are the homunculi supposed to provide an underlying causal explanation of how judgements change when subjects move forward in the forced march paradox, they are also supposed to provide an explanation when subjects move back again. When a subject classifies images along a forced march and eventually their judgement switches from one category to another—a moment of transition Raffman calls a *gestalt switch*—some of the previously encountered images will also switch to the new class if the subject were to be asked to judge them again. This phenomenon has been empirically validated and is what Raffman refers to as the ‘backward spread’ of the extension of *F* (Raffman, 2014, ch. 5). Raffman (1996) acknowledges that there remain a number of unsettled explanatory gaps in her approach. One such example is why, over multiple trials, the *gestalt switch* does not always occur at the very same location in the sequence, nor with the same length of backward spread. Raffman’s account is limited in explanatory power. Graham Priest argues that contextualism does not solve the sorites proper, as it does not, in actuality, address the nature of the cut-off. Contextualism merely tries to explain, on the basis of epistemic reasons, why the cut-off is never encountered by the observing individual. Priest writes:

> It is not clear that contextualism really does explain why we find the existence of a cut-off counter-intuitive. Suppose that we look to find the cut-off at some point or other. The similarity constraint explains why it is not there. With respect to that context it must therefore be elsewhere—outside the area of salience. But the thought that there is a sharp cut-off point somewhere else is still as puzzling as before. The tolerance of vague predicates seems, after all, to be a general phenomenon, not simply localized to the area of salience. What could make *that* the cut-off point is therefore just as puzzling as before. (Priest, 2003, p. 15)

Nonetheless, Priest expects that maybe, with the advancement of science, an appropriate explanation for the cut-off can be found.

Furthermore, Rosanna Keefe (2007) suggests that Raffman cannot explain why we are psychologically inclined to accept STP from WTP—a problem discussed in detail below. On the other hand, the account seems too restrictive regarding how categorisation occurs along the sorites sequence. It is said to be limited to the case of pairwise judgement in the very isolated case of a forced march (Keefe, 2003, 2011).

These limitations originate from Raffman’s claims about the presence of homunculi, which are not in line with contemporary views of the brain. The claim that these homunculi exist seems like an ad hoc explanation that serves the purpose of explicating a very specific phenomenon. Moreover, due to its specificity, that explanation prevents her account from being fully applicable at a more general level. In particular, the idea of a discriminator that, as detailed in the quotation above, avoids categories shifting too quickly seems to provide an explanation for why pairwise judgements of instances satisfy the WTP. However, this explanation does not help us approach general questions regarding vagueness. Keefe (2003), for instance, challenges Raffman’s point that we somehow envision ourselves walking a forced march when we are confronted with the sorites paradox. Keefe says that it might just as well be possible to believe that STP is true on the basis of abstract reasoning; or, at the very least, it seems that the properties of vagueness can be witnessed without relying on the notion of *pairwise* inspection. This is explained by the following example:

> Suppose you present someone a whole Sorites continuum and ask them to show the point of the left-most non-red patch, where they are obliged to respond by picking a patch arbitrarily. We can’t then excuse the subject because they changed score or context *between* pairs while maintaining the rule of judging pairs the same: that way out is not available with such a scenario. (2003, p. 80)

Raffman's psychological account is, therefore, limited to the case of pairwise judgement, which does not solve the sorites paradox in general, namely the sorites proper.[3] From this point of criticism, there seems to be more to the sorites than an analysis of speakers' verbal dispositions at a given position in a sorites sequence.[4]

Within the vagueness literature, we find that not only Raffman but most contextualists also rely on some form of a *similarity constraint*, from which they draw conclusions about an object's categorisation based on its similarity to adjacent objects. This helps them to formulate a version of the WTP that is then argued to be confused with the STP. Graff Fara, for instance, emphasises the significance of salient similarities:

> If two things are saliently similar, then it cannot be that one is in the extension of a vague predicate, or in its anti-extension, while the other is not. (Graff Fara, 2000, p. 57)

Similarly, Soames argues in favour of the following metalinguistic criterion, grounded in the concept of a competent observer:

> For any two patches of color x and y that are perceptually indistinguishable to competent observers under normal conditions, if someone who is presented with x characterizes the predicate *looks green* as applying to it, then that person is thereby committed to a standard that counts the predicate as applying to y as well. (Soames, 1999, p. 215)

And finally, Shapiro's version of the WTP is formulated as follows:

> Suppose that two objects *a, a'* in the field of *P* differ only marginally in the relevant respect ... Then if one competently judges *a* to have *P*, then she cannot competently judge *a'* in any other manner. (Shapiro, 2006, p. 8)

Contrary to Hume's account, which runs into immediate problems of vagueness, contextualists restrict their analysis to the similarity of elements *within* pairs of instances, limiting their accounts to *pairwise* judgements. Accordingly, contextualists assume that, with respect to a given contextual standard, adjacent distinguishable object pairs are always similar *enough* to warrant equal classification when judged together—otherwise the sequence would not qualify as a sorites sequence. Such assumptions presuppose that, at least for a given pair, the context is fixed and, by extension, so too is the criterion for what it means to be similar enough to fall under the same category. But these assumptions do not generally hold.

Minimising prediction error is what determines categorisation. Similarity comparisons between elements of a pair—or comparisons with some prototypical image in the mind—do not govern categorisation. The similarity between two instances of a pair typically *correlates* with equal classification, but it is not its *cause*. According to PP, the brain performs a similarity comparison not with stored images, but with a *self-generated* predicted image—an image that is constantly updated as the features defining a category evolve. The brain is not a photographic archive

[3] The same concern of not addressing the sorites proper can be made for other contextualist accounts such as Graff Fara's (2000) view, even if she is less explicit about the cognitive mechanisms involved.

[4] These issues are further discussed in Bones and Raffman (2019) and Hyde and Raffman (1997).

comparing incoming data with static templates from memory. Instead, it *generates* predictions about incoming sensory data, and when those predictions are inaccurate, the neural model adapts (*neural plasticity*) to improve future predictions. Moreover, the conditions under which a patch is judged to be similar *enough* to the predicted image to fall under a given category are also constantly shifting. Consequently, even for a given pair of instances, the associated categories may differ when necessary.

The similarity comparison in PP is *not* a comparison between different objects 'in the world', or objects that could potentially be seen, or images directly stored in memory, or anything alike. Rather, it is a comparison between somewhat objective sensory data—the light reflecting from an object's surface, resulting in the activation of the subject's receptors in the retina and thereafter the first cortical layer—and the subjective assumptions made by the observing, cognising system about the sensory data. Since the generated image to which a similarity is made is dependent on the task of minimising prediction error, I suggest that, in order to make sense of the sorites, we rely on the overarching principle of minimising prediction error.

This does not mean that I am completely opposed to the above views; rather, they require refinement. Consider Graff Fara (2000), who claims that extensions shift in accordance with our interests. She says that when we focus our attention on a certain position in the sorites sequence, the extension changes based on psychological factors. Even though Graff Fara has, in contrast to Raffman, no cognitive explanation to offer for these shifts, her intuitions are partly in line with PP. The brain does not build representations of the world for the purpose of being veridical; instead, it constructs categories that minimise prediction error based on our focus of interest. Making good predictions of the phenomena and their hidden causes that capture the cogniser's attention is the sole purpose behind the mechanisms underlying categorisation. The action leading to a change in focus is again explained by the purpose of minimising prediction error or surprising states; for instance, shifting focus might save you from being surprised by a lion hiding behind a bush. This change of focus triggers an adaptation of mental categories in order to prepare for what is likely to happen.[5]

According to Graff Fara, two things are of the same category when there is no additional value in discriminating between the two for a given purpose. Whenever we shift the focus of our attention, extensions of a category can also shift due to a change of interest. Two objects can, in some situations, be counted the same when they are on other occasions counted differently. Graff Fara's account lacks an explanation of *how* extensions shift (Åkerman, 2009, p. 77), is limited to pairwise judgements, and thus could not address the results of my experiment. She might provide an answer as to why two neighbouring instances are judged to be an F, but this fails to explain why it is not possible to locate the last instance of an F (Heck, 2003).

[5] Active inference is a theoretical approach to connecting action, perception, learning, and the encoding of uncertainty into one theory. It relies on the assumption that actions are made to fulfil predictions based upon inference to sensory data (Friston et al., 2016).

15.4 The Idealist Approach

Many contextualists assume that the cut-off point exists, in one way or another, independently of the subject confronted with the sorites paradox. For the epistemicist, for instance, there are objective standards for using a term, determined by the language community. The assumption that the cut-off must have some objective characteristic comes naturally: if the cut-off is epistemically hidden from us, then it appears contradictory to believe that the cut-off is part of us. But this is precisely what the idealist suggests. For an idealist, the structures we find in the world are determined by the observing subject. This does not mean that the subject is not in a back-and-forth interaction with a language community, or that the subject cannot be mistaken about how it uses language. However, the structural aspects of how we think and reason about objects are equally shared among all (healthy) people and do not reflect something in the world, but rather something about our mental processing. The fact that we distinguish between bald and non-bald men is a general fact about us being creatures who do not merely see the world as one unstructured entity, but rather as a multiplicity of individuated objects. In the case of vagueness, the idealist suggests that the cut-off must therefore be imposed by us onto the world.

According to PP, the guiding principle of the brain at the unconscious neurological level is to minimise the prediction error of incoming sensory data, given an adaptive probabilistic model. Minimising prediction error amounts to a handling of sensory data that is highly context-sensitive. Consequently, 'whatever you end up experiencing will be the result of … context-sensitive processing of sensory signals' (Kirchhoff, 2019, p. 87). This chapter argues that the principle of prediction error minimisation allows us to formulate a radically new theory of vagueness that is, in spirit, similar to previous contextualist accounts, but ultimately provides a very different way of looking at the problem.[6]

15.5 Detecting the Cut-Off Point

A central, intuitive assumption underlying all accounts of vagueness states that for a given sorites sequence running from F to $\neg F$, we have no epistemic access to the cut-off point, meaning that there exists no method by which an individual can detect a cut-off point (x_c, x_{c+1}) for which Fx_c and $\neg Fx_{c+1}$. If we had reason to reject this assumption, then the paradox would be easily resolved by a straightforward rejection of the induction step.

[6] For the sake of simplicity, this chapter focuses on the sorites paradox of perceptual objects and does not discuss any sequence involving abstract terms such as 'freedom' and 'captivity'. Once we have examined how language is processed from a PP perspective in a later chapter, the account can be interpreted as providing a solution to abstract terms as well, and I leave it to the reader to do so.

But what does it mean to be able to detect the cut-off of a sorites sequence? Imagine a *forced march* where a subject is asked to judge each instance it encounters along a sorites sequence as to whether it is an F or not. By the very nature of the situation, the subject will at some point inevitably transition from judging instances as Fs to judging them as not-Fs. The moment of transition is what Raffman (1994) calls a *gestalt switch*. Did the subject in this very moment detect the cut-off of the sequence? Of course not. For two points along the sequence to be considered as a cut-off, the subject must *reliably* judge the first instance to be F and the second instance to be not-F. I will refer to this as the *reliability criterion* of the cut-off point. In the typical case of a sorites sequence, this criterion is not fulfilled, meaning that the subject will not consistently refer to the same point as the cut-off, neither during the forced march nor when no longer in the forced march situation.

The lack of reliability is exemplified in an experiment discussed in detail by Raffman (2014, ch. 5). In this experiment, subjects were presented with images they had to classify along a sorites sequence running from F to $\neg F$. Once a gestalt switch took place, the direction in which images were presented was reversed, and subjects were presented with images running from $\neg F$ to F, starting from where they had left off, that is, where the gestalt switch had occurred. What the experiment shows is that previous instances in the sorites sequence are judged *differently* after a gestalt switch. For example, if a subject judges the person x_0 as bald, ..., x_{30} as bald, x_{31} as bald, and x_{32} as not bald, then the subject will typically in the next moment judge x_{31} and then x_{30} also as not bald, despite having said otherwise a few moments earlier. This phenomenon is what Raffman refers to as the *backward spread* of the extension of F, which clearly indicates that during a forced march situation a cut-off is not found, since the subjects contradict themselves and do not reliably identify one point as the point of transition.

However, the reverse is also true: if a subject behaves consistently—meaning that no backward spread can be observed, and the subject repeatedly asserts that x_{31} is bald and x_{32} is not bald, even when no longer in a forced march situation and having been exposed to other instances—then one can say that the exact cut-off of the sequence has been identified. We would typically conclude that in such a case the sequence was never a sorites sequence in the first place because the sequence's *granularity*, meaning the difference in change from one instance to the next, was too large.

Those who have explained the sorites paradox to friends or students might be familiar with the situation where somebody initially does not understand the paradox. The friend or student might state that three stones do not constitute a heap, but four stones do, suggesting that, along this given sequence, one additional stone really does make the difference. In such a case, the easiest way to persuade the other that there is a problem at hand is by presenting a more fine-grained sequence. For instance, instead of adding entire stones, one might ask whether three stones plus a grain of sand constitute a heap. Eventually, the friend or student will understand the problem—but only by identifying a suitable sequence where they have not detected a cut-off point. In order to contrast sequences that generate paradoxical intuitions from others, I define a *morphing sequence* to be *any* ordered sequence of instances with distinguishable, incremental changes from x_0 to x_n, whereby Fx_0 and $\neg Fx_n$. A morphing sequence can have a cut-off point that is apparent to anyone. In the literature, the sorites paradox and the concept of vagueness are always explained by providing examples of some morphing sequence at a fine-enough granularity. The morphing sequence only then appears to us as a *sorites sequence* when we have not identified a cut-off point at a given time.

It seems intuitively true that any non-soritical morphing sequence can be turned into a sorites sequence when the level of granularity is decreased, that is, becomes more fine-grained. What also appears intuitively true is that a sorites sequence cannot become a non-soritical morphing sequence without imposing additional, explicit defining conditions, such as: 'a pile of 10 stones is a heap, and a pile of 9 stones is not'. From these observations, let us provide two definitions of what a sorites sequence is:

Sorites-D1 A morphing sequence for which there exists no cut-off, or for which we have no epistemic access to the cut-off.

Sorites-D2 A morphing sequence for which we are pre-theoretically inclined to accept the induction step of the sorites paradox.

The existing literature generally does not offer a definition of the sorites but only relies on examples of morphing sequences. Further, it does not draw this distinction between Sorites-D1 and Sorites-D2 because the two definitions appear to have the same extension. It seems that whenever a sequence is a Sorites-D1 sequence, it is also a Sorites-D2 sequence, and vice versa. Likewise, any theory to date on the sorites sequence makes the following two assumptions:

No access assumption For any morphing sequence, there is some level of granularity at which there exists no cut-off, or for which we have no epistemic access to the cut-off.

Inclination assumption For any morphing sequence, there is some level of granularity at which we are pre-theoretically inclined to accept the induction step of the sorites paradox.

In the following, I will show that any morphing sequence of distinguishable objects that is soritical according to Sorites-D1 can be turned into a non-soritical morphing sequence. Thus, the *No access assumption* is, in a certain sense, false. However, the *Inclination assumption* explains why we find the major premise of the sorites so compelling. The argument is motivated by experimental findings and then clarified and reasoned on the basis of PP theory.

15.6 Modified Forced March Experiment

Imagine we have a morphing sequence from black to grey, such as the one presented at the top of Fig. 15.2. Most people would, at first glance, find the change between each of the coloured patches marginal enough for the sequence to be considered a

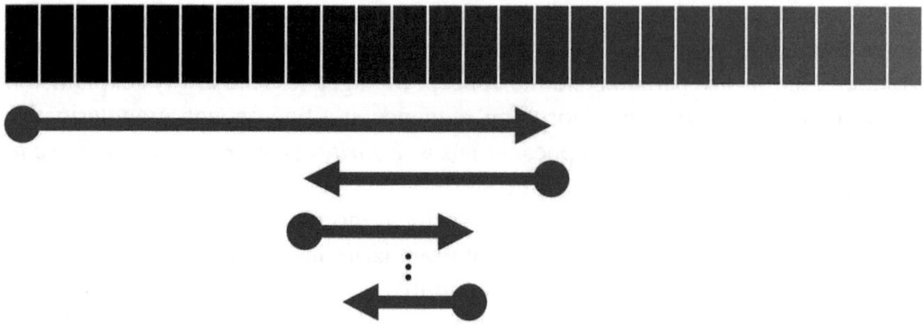

Fig. 15.2 Morphing sequence from black to grey with a forced march that switches direction whenever the subject changes their judgement from 'black' to 'not black' or vice versa

sorites sequence. As expected, we find the phenomenon of a backward spread when performing a forced march experiment with reversed patch order once the observing subject transitions from 'black' to 'not black'.

Whenever a subject is moving along a forced march, it would be wrong to infer that the subject has *found* the cut-off point after changing their judgement from 'black' to 'not black'. This is because of the occurrence of backward spreading. The presence of backward spreading shows that the subject's judgements are unreliable and inconsistent, implying that no cut-off was found.

But what happens when we repeat this process of changing the direction of the forced march every time the subject changes their mind about the category (see Fig. 15.2)? Does the subject always exhibit an equally long backward spread each time the direction of the forced march is reversed? Or how does the backward spread change over the course of the experiment? Performing such an experiment shows that the length of the backward spread tends to decrease over time. After only a few directional changes, many of the subjects begin to alternate between two patches that are right next to each other. This is shown in Fig. 15.3, where we see the length of the backward spread for 100 participants who were divided into two groups and had an exposure time of 500 ms and 3 s, respectively, for each patch they had to judge.[7] Furthermore, when the subjects alternated between two consecutive images and were then distracted by being asked to judge patches of prototypical black, 23% of them were still able to rediscover the position along the morphing sequence where they had transitioned from black to not black. This changes our interpretation of whether the subject managed to detect the cut-off point. When a subject makes alternating judgements between two adjacent instances without any backward spreading, and even remains consistent when not in a forced march situation, then we are no longer in a position to claim that the discrete cut-off has

[7] The survey was built using PsyToolkit (Stoet, 2017) and was distributed via the crowdsourcing website prolific.co to English-speaking adults.

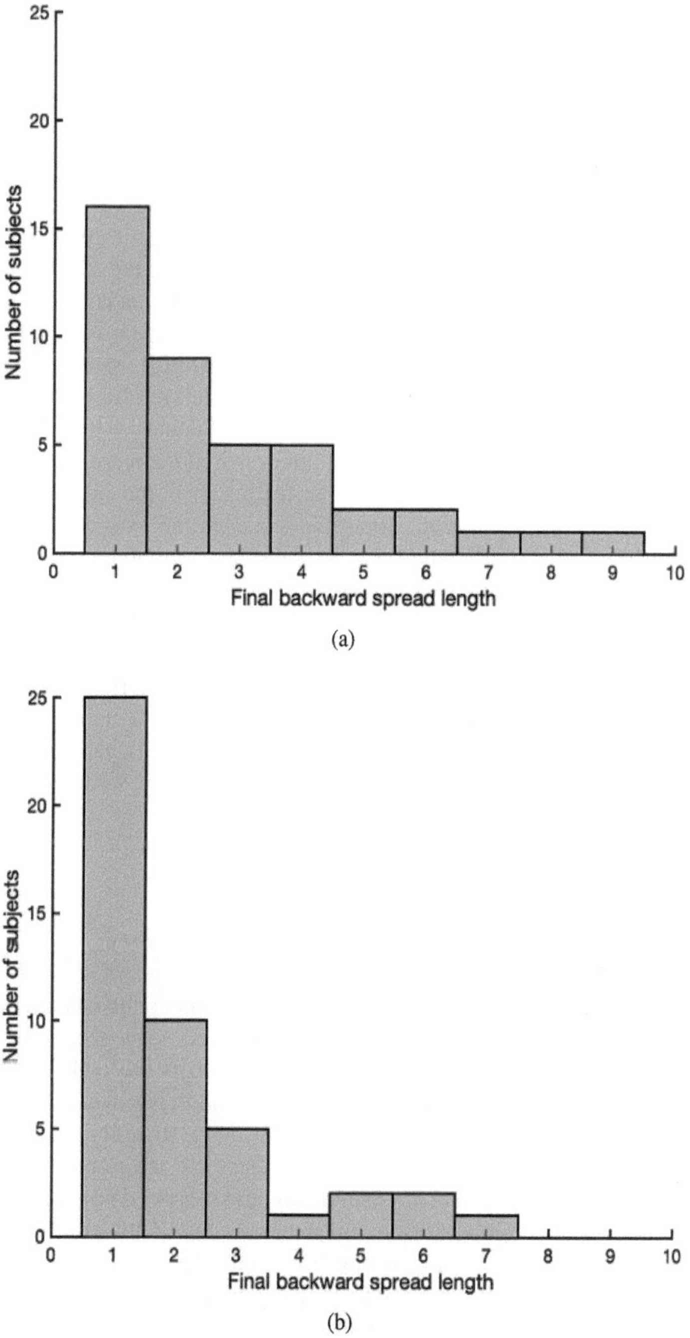

Fig. 15.3 Histogram of the minimal backward spread length for an image exposure time of (**a**) 500 ms and (**b**) 3 s

not been found. Thus, in the presented experiments, some subjects should be said to have found a cut-off along the given sequence.

These results are, to some extent, counterintuitive, as when one looks at the given sequence running from black to grey, one would hardly ever say that there can be a cut-off point. It is perhaps for this reason that all existing theories about vagueness have always taken a sorites sequence as a given, rather than a morphing sequence that can dynamically change from being a soritical sequence to being a non-soritical sequence. While the results of the experiment could arguably be given various interpretations when considered in isolation, the interpretation one can give them from a PP perspective is very straightforward, as is the interpretation from the idealist perspective. Furthermore, these experimental results are largely for illustrative purposes, showing how the theory of cognitive learning reveals itself in real life. Even if the data might not be fully conclusive, the theory of PP can independently be interpreted to suggest what is argued below.

For the realist, the facts about how cuts are drawn along a sorites sequence are interpreted to be external to the mind. It is, for instance, the language discourse that determines at which location along the sequence the *true* cut-off point lies. Accordingly, the realist will likely deny that the subject has actually found the cut-off point, but will have much explaining to do as to why the speaker community has ultimate authority over all possible sorites sequences. The idealist, on the other hand, is in a much more comfortable position. For her, vagueness is a structural property we impose onto the objects we encounter, and there are certain ways in which relevant mental mechanisms make us experience a morphing sequence as soritical at some times and not at others. To comprehend what is precisely happening, we must look deeper into what the brain is doing and let PP tell us how we must interpret the empirical results.

15.7 Cognitive Learning

Despite people being able to find cut-off points, it remains the case that a cut-off cannot be found immediately upon inspection of a sorites sequence. So why did the process of moving back and forth along the sorites sequence make a difference? What is going on at the unconscious level is that the brain is *learning*, meaning it is adapting its internal representations by neural plasticity. Roughly speaking, during the forced march, unconscious learning mechanisms are triggered since the brain cannot deal with the sensory data in a useful way, and sensory data is only useful if it can be grouped and distinguished from other sensory data. Thus, the brain 'realises' that it can only cope with the situation well if it introduces a new representational distinction. After moving back and forth along the morphing sequence, the cognitive system eventually draws a clear-cut distinction where it finds a cut-off most useful. The outcome of the second stage of the experiment shows that many participants fail to reidentify the region of the cut-off point due to the short exposure time of a determinate case of black. More practice would be needed to make the same

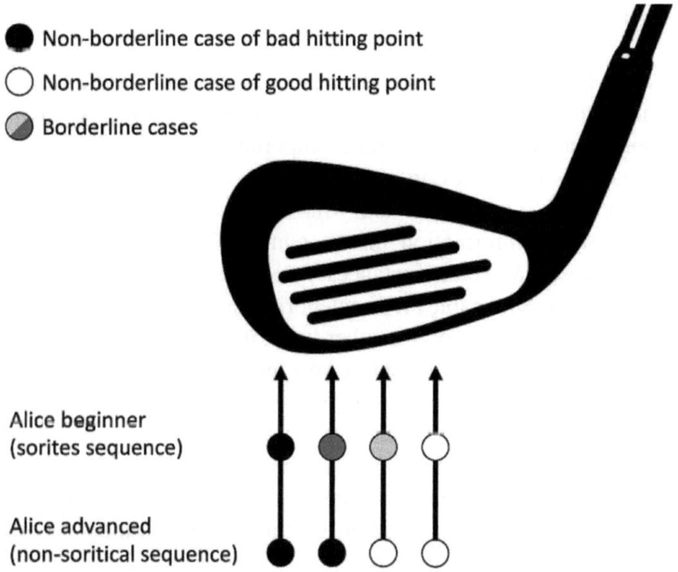

Fig. 15.4 When Alice begins to learn golf, one can construct a sorites sequence between a good and a bad point of contact between the club and the ball. This same sequence will turn out to have a clear cut-off point once Alice has become more advanced

discrimination repeatedly, and such practice is to some degree afforded by higher exposure times. Furthermore, once the subject's brain encounters data from more stereotypical cases of black, it may no longer 'feel the need' to be able to make more precise discriminations at the boundary.

Constructing new cut-off points is what unconscious cognitive learning often amounts to. When the brain is provided with sensory data for which its current representational structure has no useful cut-off points defined, it adapts its representation. This can best be illustrated by turning to the way we ordinarily learn and progress in a domain where learning is not achieved by linguistic means or by people verbally telling us what is right and what is not.

Imagine Alice, who passionately starts learning golf. At the beginning of learning how to play, she cannot distinguish a 'good' from a 'bad' point of contact between the ball and the golf club. For her, it appears as if half an inch more to the front or to the back of the club would not make any difference in how the ball will fly (see Fig. 15.4). However, becoming more advanced in the sport and repeatedly playing similar shots over and over again, Alice eventually learns to distinguish a good from a bad shot by detecting the smallest differences between the two. This example illustrates how our brain introduces increasingly fine-grained distinctions when these are useful, resulting in finding a cut-off point for a given sorites sequence. When Alice was a beginner, a sequence running from 'good' to 'bad' at different golf club locations would not have had a clear cut-off point. However, once she trains a lot, a cut-off point appears to her for this given sequence (see Fig. 15.4, bottom).[8]

Humans are finite beings with finite resources to form neural encodings that compress information efficiently. Storing information efficiently is only possible by having a sophisticated way of filtering unimportant information from information that is expected to be useful in the long run. Vagueness illustrates how this compression process takes place and what the brain is doing when we direct our attention towards information that is commonly conceived as unimportant. Non-borderline cases are data instances for which the brain has developed a useful representation that enables it to categorise data efficiently for making useful predictions. Borderline cases, on the other hand, are those data instances for which the brain has not found it worthwhile to invest resources in creating a representation in order to make successful predictions of this type of data. It simply has, up to this point, not been interpreted as relevant. However, when we direct our attention to borderline cases, they become relevant simply by the fact that we are giving them our attention, and the brain immediately starts negotiating different categories against each other, determining whether existing categories must be broadened in their extension or whether new categories must be established.

In the case of our experiment, in which a morphing sequence runs from black to grey, the brains of the participants are at first not processing the data consistently, resulting in a backward spread. This inconsistency is, however, biologically and for survival purposes not useful, and the brains of the participants develop more consistent representational frameworks in order to deal with the newly encountered data more effectively. This ultimately results in very low to zero backward spread, increasing the consistency of the participants.

15.8 The Nature of Vagueness

Contextualism has been criticised for lacking generality. Keefe (2007), for instance, says that she '[s]urely . . . can just consider . . . [the induction step] in the abstract in a single context' (p. 284), and it seems evident to her that our attitude towards the induction step does not originate from the inspection of different pairs. There must be, in her view, a different reason as to why we are inclined to trust the major premise.

Having understood that cognition adapts to a morphing sequence by examining it, it is possible to provide a general account of vagueness in the contextualist spirit. The sorites paradox is a consequence of the brain not yet having had the time to adapt its representations to newly recognised relevant data. I believe that the

[8] In the case of hitting a ball with a golf club, we of course have no *given* discrete sequence but rather a continuous one. Nonetheless, the analogy is valid if we imagine defining a sequence horizontally along the club.

contextualist strategy of pointing at a confusion of ours is an adequate reply to the paradox but requires refinement. The error we make is that we think that STP is true:

STP If a sensory datum x_i looks like an F, then the adjacent datum x_{i+1} along a sorites sequence also looks like an F.

However, this premise of the sorites induction step must be replaced with the following:

WTP* If a sensory datum x_i looks like an F, then there exists *some* morphing sequence for which the adjacent datum $x_{i+1/2} < x_{i+1}$ also looks like an F.

I will first explain why WTP* is a possible general reply to the *psychological question* and then why, in contrast to STP, it avoids the sorites and thus also serves as a possible reply to the *logical question*—the question of how the sorites paradox is logically resolved. I am not saying that it is the only possible reply one can give to the *psychological question*, as I do not see a reason why multiple factors could not be involved when we form our judgements about the major premise. But I believe it is a reply that is most closely connected to how cognitive processing in creatures like us works, and to the causes of vagueness rooted at the most fundamental level. It is furthermore a reply that is general and not dependent on examining a specific sorites sequence, and is universal in the sense of applying to all humans with a healthily functioning cognitive system.

The rough idea behind the PP-inspired account, motivating WTP*, is the following: when we are presented with a morphing sequence, we experience it as a sorites sequence as long as the brain has not yet learned the appropriate categories to draw a sharp cut-off point. Conversely, when we do find a cut-off point, the brain has established a useful representation for the data. In theory, for any morphing sequence with distinguishable patches, one could learn to identify a cut-off point. The reason, however, why we find the sorites' major premise so compelling is that there is always a way to make a given morphing sequence even more fine-grained, such that it turns back into a sorites sequence of data instances we have not yet learned to distinguish. Let me explain this in more detail.

While subjects can always detect a cut-off for a given morphing sequence when using the right method, there is always some level of granularity to which the sequence can be decreased such that the subject, at that given moment in time, does not see a cut-off. I illustrated this with the example of a student to whom we explain the paradox by decreasing the granularity of the presented example. WTP* formalises this idea by stating that we can always find *some* morphing sequence that initially appears as a sorites sequence. Using the right method, this new sequence will eventually no longer be soritical, but once granularity is further decreased, it will again appear soritical, and so on. WTP* therefore expresses that there is always *some* level of fineness for which our brain has not yet learned how to make useful categories. It is this fact (among possibly others) that makes the major premise of the sorites so compelling: there will always be *some* $x_{i+1/2}$ in *some* morphing sequence that is an F if x_i is an F.

Assuming that there exists a $x_{i+1/2}$ between x_i and x_{i+1} such that Fx_i, $Fx_{i+1/2}$, and $\neg Fx_{i+1}$, we are in a position to answer the *psychological question*, since we can always find an incremental change to some x_i that is closer to $\neg F$ while still being an F. This helps motivate the induction step when philosophers present the sorites paradox: one can pick an arbitrary position along a sequence and identify ever-increasing values that still count as Fs, despite approaching the direction of a $\neg F$ instance. Philosophers support this by increasing the granularity of the morphing sequence as needed. The explanation of this intuitive presupposition—that there exists such an $x_{i+1/2}$—is addressed in the next section. For now, we can see how WTP* provides answers not only to the *psychological question* but also to the *logical question*.

In contrast to STP, WTP* only states that *some* morphing sequence exists for which $x_{i+1/2}$ also looks like an F. From a logical point of view, this modification protects WTP* from necessarily leading to a paradox. To see this, consider the mathematical series of the reciprocals of triangular numbers: $\frac{1}{1} + \frac{1}{3} + \frac{1}{6} + \frac{1}{10} + \ldots$. Contrary to some intuitions, this series does not diverge to infinity despite adding an infinite number of values, but instead converges to 2. Thus, merely the fact that one is increasingly moving into a certain direction, even for an infinite number of steps, does not guarantee that one is moving beyond a certain point. Coming back to WTP*: even if there is always some adjacent datum that looks like being of the same kind as the one before, this does not mean that we are actually heading towards an absurd conclusion or even towards non-borderline cases. If, for a given morphing sequence, we look at a non-borderline x_i, then there is some sequence x_{i+q_1}, x_{i+q_2}, x_{i+q_3}, etc., that we could add between x_i and x_{i+1} such that the series never reaches a borderline instance (whereby q_i are just arbitrary fractional indexes for distinction purposes). We could thus construct a new morphing sequence $[x_1, x_2, \ldots, x_i, x_{i+q_1}, x_{i+q_2}, x_{i+q_3}, \ldots, x_{i+q_m}, x_{i+1}, \ldots, x_n]$ with arbitrarily large m, for which x_1 to x_{i+q_m} are non-borderline Fs. We can therefore accept WTP* as a logical premise that answers the *psychological question* without necessarily running into a contradiction. WTP* can be modified to WTP**, which avoids potential contradictions completely by only considering the subset of morphing sequences that are non-soritical:

WTP** If a sensory datum x_i looks like an F, then there exists some non-soritical morphing sequence for which the adjacent datum $x_{i+1/2}$ also looks like an F.

Both WTP* and WTP** are true and reflect our motivation for why we are inclined to agree with the sorites induction step while avoiding the paradox from a logical perspective. I conclude that WTP* and WTP** provide answers to both the *psychological question* and the *logical question*.

15.9 Robust Cognitive Systems

From what has been discussed so far, there might be two open questions. First, one might ask why it is the case that there is always some $x_{i+1/2}$ for which $Fx_{i+1/2}$ and Fx_i. Second, I have shown that it is always possible to construct a soritical morphing sequence, and this has been argued to explain how sorites sequences come about. However, it might still be unclear how we run into a paradox for a *given* sequence. These two questions are closely related, but after the preceding analysis they are no longer of a primarily philosophical nature. The first question is answered by pointing to the specific quality of *robustness* in our cognitive processing. The second is addressed by the contextualist view, according to which we experience surrounding instances along a given sorites sequence as belonging to the same category. My account, however, extends on this, by explaining how an exact cut-off point eventually emerges through a learning process. For these details, let us take a step back and look more closely at how human cognition operates when attempting to classify instances along a sorites sequence.

We have learned that, instead of relying on naïve classification, the brain operates differently. It is hierarchically structured, with higher cortical regions actively generating percepts and imposing categories onto the sensory data. These regions attempt to predict the sensory input in a *robust* manner by relying on prior probabilities and likelihood estimates. To determine the category of a given sensory input, the brain selects the one that minimises average prediction error and adjusts its categories when necessary. In simple terms, the brain does not passively assign a category based on immediate perception; rather, it selects the category that is most likely to reconstruct the data both at the present moment and in the near future, based on what it has previously encountered.

For predictions to be successful on average, categories are selected and adapted in a robust manner. This means that slight variations in the input data will only result in categoric differences if necessary (i.e., if the probabilistic success exceeds a certain *threshold*). Wherever possible, such differences are avoided through slight refinements of the category. For instance, when an observer moves along a sorites sequence, certain non-borderline cases are easily predicted as Fs. However, as one moves away from these non-borderline cases, the system adapts by extending the extension of category F in order to resist unnecessary categoric changes and thus maintain robustness.[9]

This is illustrated in Fig. 15.5, where instances such as x_1 and x_2 are likely to be associated with category F. However, when the subject focuses on, say, x_4, the cognitive system predicts that it is also likely for the subject to later focus on x_5 and perhaps x_6. To prevent a categoric change, the system may slightly adapt F into F^*

[9] The more categories a cogniser possesses, the smaller they are in extension and the less robust these categories are to noise. This is why the brain avoids introducing new categories unless necessary.

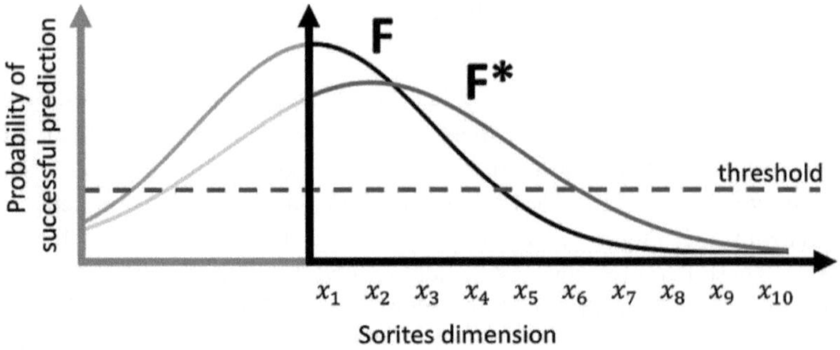

Fig. 15.5 A category F (black line) is altered to F^* (dark grey line). This results in a trade-off between making worse predictions for determinate cases and better predictions for boundary cases, which can be valuable to avoid a drastic category change

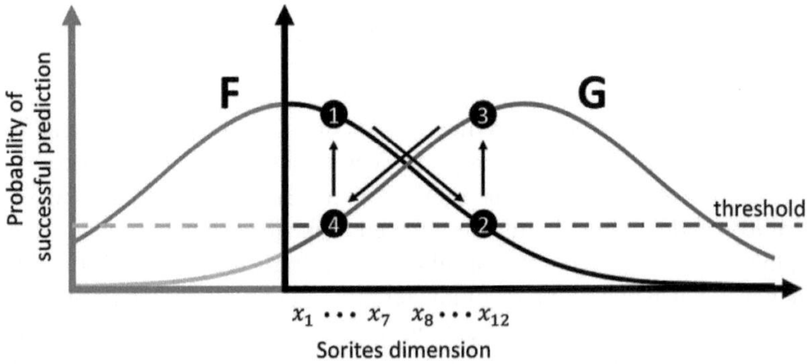

Fig. 15.6 Four stages during a direction-changing forced march: (1) determinate state (high certainty), (2) before gestalt shift (low certainty), (3) after gestalt shift (high certainty), (4) before gestalt shift (low certainty)

without actually changing the category. Categoric changes are triggered only when absolutely necessary, as they compromise robustness. Such a change is initiated when a certain threshold of unlikely successful predictions is exceeded. In a case where we have a categoric change, such as during a gestalt switch, instances of focus can no longer be properly predicted. Any further extension of category F would come at the expense of a different category G, which could undermine long-term predictive success. This is illustrated in Fig. 15.6, where the subject moves along the sorites sequence until predictions made using category F are no longer likely to be successful. Once the probability falls below a certain threshold, a category change to G occurs (gestalt switch). If the subject then proceeds in the opposite direction, the same change takes place again, and the cognitive system returns to predicting instances with category F.

Moving back and forth along the sequence triggers cognitive learning. This process allows the system to develop categories that are more robust to future

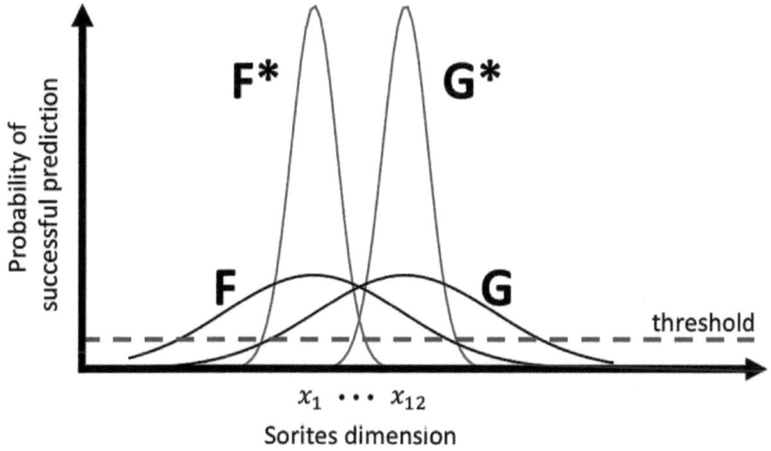

Fig. 15.7 Earlier categories (black lines) are adapted categories (grey lines) due to learning. Learning narrows the curve of the generating probability distributions for making better predictions by achieving a very low intersection point that is close to the threshold level. Consequently, there is a sharp cut-off point along the sorites sequence

inputs and avoid unexpected category switches caused by reaching a low-probability threshold. The result of this adaptation is illustrated in Fig. 15.7, where we see that the system has been updated so that none of the instances it has encountered produce prediction probabilities that conflict with another class. This refinement can lead to the emergence of a precise cut-off point.

Robustness in PP entails that for any object x, there will be some nearby object x_ϵ that is or would be categorised in the same way *at that moment*. If no such object existed, the system would not meet the conditions for robustness. This explains why there is always some $x_{i+1/2}$ such that both Fx_i and $Fx_{i+1/2}$ hold.

Robustness also clarifies how it is possible to experience pairs of instances as belonging to the same class, while still eventually arriving at a situation in which an exact cut-off is discovered along the sequence. By definition, a sorites sequence is one for which the system has not yet learned to detect a cut-off, and the instances are close enough to trigger gestalt switches. The system has not yet reached a robust state but is likely to do so if the subject behaves in a manner that leads the system to develop a useful cut-off. For a given sorites sequence, we encounter a paradox precisely because we have not yet adjusted our internal categories in a way that allows the cognitive system to predict instances along the sequence in a robust fashion.

Summary

Learning theory and PP show that our ordinary philosophical intuitions concerning the epistemic inaccessibility of the cut-off point along a sorites sequence are, in some sense, mistaken. This fact cannot be accounted for by any widely recognised theory of vagueness to date, as these theories underestimate the significance of cognitive learning mechanisms and diachronic processes of categorisation. PP suggests that the brain is highly context-sensitive and operates on the principle of minimising prediction errors. The theoretical framework of PP allows for a refined contextualist approach that can account for these new findings while also addressing the general limitations of contextualism.

Chapter 16
Information Processing and the Clash

While vagueness is omnipresent in the manifest image, it is much harder to find in fundamental physics. Adding one atom to another results in two atoms, not one. The reason for this is that at this level of precision, we have complete information about the number of objects detected, for instance, in a scanning tunnelling microscope or a particle collider.

In the last chapter, I argued that vagueness arises from the absence of a learned structure that would allow us to adequately represent certain data. We saw that in the manifest image, vagueness persists even when increasingly precise representations of data are available, because there is always some data that has not yet been adequately learned. This is no longer the case at the most fundamental level, where scientific instruments are used to detect objects. This results in a structural clash between entities in the manifest image and scientific entities.

I have argued that we can make sense of our manifest ontology by turning to the observer and using PP and the FEP as conceptual frameworks. From these general frameworks, we have seen that we can make sense of the structures we ordinarily observe. What I have not yet shown is how the different levels of reality are related. Can we account for the structural differences between these levels within a single framework? In this chapter, I will explain how entities at different length scales relate to one another, thereby offering a solution to the puzzle of how inconsistent structural realities at different scales can co-exist while remaining interdependent.

The idea behind the proposed account is roughly as follows: to individuate objects or patterns in (sensory) data, a cogniser must always impose some prior knowledge or information onto the system it is trying to cognise. This information differs between various scientific practices, as well as in cases where the cogniser is not engaged in scientific activity. In ordinary experience, what we perceive depends on the principles by which our cognitive system operates and on the information it brings to bear during perception. When we engage in empirical scientific research, the entities we observe result from our cognition being *extended* by an experimental apparatus. This extension can, in principle, be expressed in information-theoretic

A. C. Schwaninger, *Cognitive Metaphysics*, Synthese Library 529,
https://doi.org/10.1007/978-3-032-05420-3_16

terms, since the experimental setup itself adds and modifies the information entering the observed system. This means that the objects we detect in experiments resemble, in an information-theoretic sense, the means by which they are detected. The relationship between objects in the extended and unextended states appears opaque in practice, since quantifying the prior knowledge embedded in an experimental device is difficult. It is this opaqueness that gives rise to the structural clash between different images of the world.

I will first discuss the puzzles arising from Sellars' dichotomy and the problems associated with existing proposed solutions. I will then present my own alternative approach by focusing on the cognitive state of the observer. This approach is illustrated with an example in which a clash between two descriptions clearly results from the observer's information import. A detailed response to Sellars' challenge is provided at the end of the chapter.

16.1 Images of the World as Different Linguistic Frameworks

We have seen that the eliminativist dismisses any judgements that are not compatible with the scientific image. This includes, arguably, our ordinary judgements about causal events (Aronson, 1971; Dowe, 2000), intentional judgements (Churchland, 1981; Kornblith, 2002), and secondary quality judgements such as 'the sky is blue' (Maund, 2011), as they are all inconsistent with the fundamental sciences.

The challenge of the *clash* can be broken down into several different aspects. First, since reality presents itself in scientific practice as being divided into layers, and accordingly many philosophers assume a scale-relative ontology,[1] the ontological underpinning of the scale *itself* must be explained or included in an all-encompassing ontology of the world. The challenge is then to offer an adequate ontology that provides not only the resources to explain 'how to reconcile the ultimate homogeneity of the manifest image with the ultimate non-homogeneity of the system of scientific objects' (Sellars, 1963, p. 36) but also how the different domains of science hang together and the *reasons* why they appear disconnected.

Second, scientific practice is tied to the manifest image since experiments always take place *within* the manifest image. This is sometimes said to challenge Sellars' dichotomy: 'Sellars apparently overlooks ... that science is fundamentally a *practice* conducted from within the manifest image' (Macarthur, 2017, p. 65). On these grounds, a liberal naturalist like David Macarthur is led to think that there is

[1] As discussed in earlier chapters, this view is prominently supported in Ladyman and Ross (2007), who claim that the objects of existence are roughly what Dennett (1991) termed real patterns: relatively stable and enduring structural patterns within the data. These patterns exist at their own scale of measurement, resulting in a layered conception of reality. No insight, however, is given as to what manifests the scale itself in ontological terms.

no real distinction between the images, and the existence of a clash between images must therefore be denied. It is, however, doubtful that this view actually does justice to Sellars' original problem, since there clearly seem to be inconsistencies at hand. Nevertheless, the point raised creates a challenge for all those accepting the clash: an explanation is demanded of how the manifest image can be the starting point for all scientific practice while at the same time proving to be inconsistent with scientific findings.

Third, history shows that a unification of the sciences, as dreamt of by the Vienna Circle, is unfeasible. The challenge is to explain why it is so difficult or practically impossible to reduce different levels of reality to one another, even though they are supposed to be part of a single reality.

To meet these challenges, I suggest that we should turn to what we might call the *contextualist* approach. Traditionally, this approach argues that the linguistic framework we employ shifts from one situation to the next. Within one context a speaker's judgment might be true, and in another context the judgment might be senseless or appear foolish. For instance, according to Amy Thomasson (2007, ch. 7), reference to material objects is established by a categorical framework that differs from one image of the world to the next: 'scientific theories ... do not use sortals such as "table", and if science and common sense are using sortals of different categories, the "things" picked out by the two descriptions cannot be identical' (Thomasson, 2007, p. 142). Thus, there only appears to be a clash or inconsistency when we mistakenly conjoin the distinct ontologies of different images. We can refer to this view as *linguistic contextualism*.

As I have already mentioned earlier, Thomasson has been criticised by French for anti-realist tendencies, since existence claims are said to be sortal-relative and relative to a specified domain of quantification. The contextualist's concern is that there is always a need for some linguistic framework in order to make any judgment about the world. The need for a linguistic framework, as well as the incompatibility of different frameworks, leads to the incompatibility of the different images. Robert Smithson, another contextualist who calls himself an idealist, can in some respects be compared to Thomasson, as he also thinks that the difference between the images is a difference in linguistic context:

> We have an antinomy. On the one hand, there seem to be cases where the scientific and manifest images are incompatible. On the other hand, there does not seem to be anything defective about our manifest discourse. The idealist respects both aspects of this antinomy by distinguishing the norms operative in manifest and scientific discourse. (Smithson, 2019, p. 519)

Nevertheless, French's critique does challenge the motivation to adopt the contextualist's mode of thinking. It is not merely our linguistic insufficiencies that build a contrast between the subatomic and the macroscopic world, but rather our ordinary phenomenal experiences that are incompatible with experimental findings. Therefore, it is questionable whether the analysis of language is the correct approach, rather than engaging with the principles of neuroscience and grounding those in physics. Accordingly, it remains unclear why the different images come about in the first place and why it is not possible to find a *single* context or linguistic framework that allows us to unify the different images. To postulate that there are different contexts determining the different levels of reality we encounter in the different images is merely to suspend the problem at hand. The challenge remains for the contextualist to explain why the different contexts exist in the first place. This is particularly challenging since the explanation must also provide details on why there are different linguistic frameworks *despite* reality being in some sense unified.

16.2 Cognitive Contextualism

Having presented the challenges resulting from Sellars' dichotomy as well as the problems associated with the linguistic contextualist approach, I will now offer an account that is contextualist in nature but manages to meet the existing difficulties of the traditional approach. I refer to the approach taken in this book as *cognitive contextualism*, which diverges in methodology from existing contextualist approaches by focusing on the information-theoretic, cognitive underpinnings of the observing subject or system rather than on his or her use of language.

It should be noted that the account presented here does not explain all inconsistencies between different entities at different levels of reality, but focuses only on the *structural* clashes we find between material objects and how they relate to one another.[2] Besides the example from thermodynamics discussed below, structural inconsistencies include the inseparability of entangled objects in quantum mechanics, which contrast with our ordinary experience of localised and separably identifiable individuals. I will not, however, have anything to say about other qualities that seem to cause inconsistencies, such as the colouredness of ordinary objects in contrast to the colourless entities in physics. This chapter also does not suggest that it identifies the only possible cause of structural clashes. It does, however, identify the necessity of such clashes between different levels of science.

Figure 16.1 visualises the notation used. Manifest material objects X_M (e.g., a tree or a table) are detected by a subject S by means of its cognitive system C_M. Objects X_S (e.g., a cell or an atom) that are empirically detected in some scientific enterprise require, for their detection, a method Q employed by S. This method comes in the form of prior conceptualisation and experimental procedures.[3] To be precise, the detected objects X_M and X_S are patterns in the data (i.e., non-

[2] The notion of 'structure' is sometimes said to be fundamental, thus having no actual definition. The investigation of structures is said to lie at the heart of metaphysical endeavour (Sider, 2011).

[3] The account presented here focuses on objects that are empirically found and might not fully generalise to entities postulated by pure theoretical reflection for which no empirical evidence exists (e.g., parallel universes or strings).

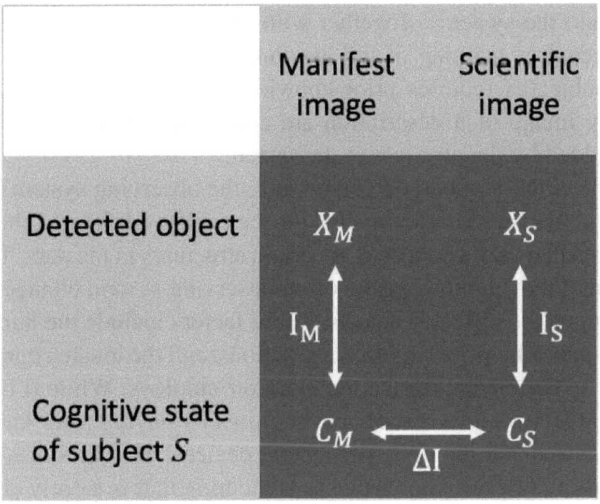

Fig. 16.1 Visualisation of notation

random and therefore compressible structures), and the method Q is a procedure that individuates these patterns.[4]

The individuation and identification of any object X requires a certain amount of information import I by the cognising system. This is explained as follows. In order to detect a pattern in any empirically acquired data, the individuation of a pattern is required, and for an observer to individuate such a structure, learning as it is characterised by machine learning is necessary. This is because a pattern is only then individuated and cognised as an entity once it is re-identifiable to the observer, which again requires the cognising process to have learned a generalised structure from the empirically collected data. To identify such a structure in the data, certain generalised features of the structure must be learned. Otherwise, every new instance in the data would appear to be unique and disconnected from other instances. We can therefore say that the observation of any object X by a cognising subject S requires a learning process.

According to learning theory, 'the incorporation of prior knowledge, biasing the learning process, is inevitable for the success of learning algorithms' (Shalev-Shwartz & Ben-David, 2014, p. 3). In other words, all learning involves prior knowledge, which means that the learning mechanisms of the cogniser must already be sensitive to detecting the features of a pattern in order to detect the pattern. Observation is thus never neutral and always requires the observer to add some

[4] The patterns under discussion here are 'real patterns', which are relatively stable and enduring structural patterns in the data that are non-random and thus allow for more efficient descriptions than random noise (Dennett, 1991). The entities detected by scientists are often said to be precisely such patterns (Andersen, 2017; Ladyman & Ross, 2007; McAllister, 2011).

information into the system.[5] Together with the previous fact that the observation of an object requires learning, it follows that the observation of the objects X by a cognising subject S requires prior knowledge of X. In summary: the observed objects in any image of a description are convoluted by the prior knowledge or information added by the observer to the system.

From an information theoretic perspective, the observing system is not only the human subject but also the experimental device employed. Jointly, they form a prior by which the system as a whole detects certain structures in the data. There are many factors by which the cognitive priors of an observing system change, and with this also the structures the observer detects. These factors include the human observer's mental state, shaped by prior experiences, actions, and the innate cognitive structure, as well as the experimental device the observer employs. While it is clear that the human mind processes information and cognises by having prior knowledge about the cognised system, so too do experimental devices. This becomes apparent once one does not think of prior knowledge as something that is merely abstract. Rather, in a materialistic worldview, it is the physical body of the observing device that entails prior knowledge about the system it can observe. Its physical realisation is not neutral in information theoretic terms. Rather, the way the device is built carries certain information about the structures it detects. The information processing of experimental devices can be complex, as in the case of a particle collider, or can take a very primitive form. For instance, filtration is a physical process that separates oversized particles from smaller ones. The filter thus imposes a certain information content onto the substance under observation, drawing a structural distinction on the basis of size.

In summary: when a subject S is conducting an experiment to detect X_S by using an experimental setup E and a method M that involves certain actions, the observing cognitive state in information theoretic terms is no longer C_M but rather the cognitive system is *extended* by $E+M$, leading to the state $C_S = C_M+E+M$.[6]

[5] Every learning algorithm makes prior assumptions about the problem at hand. This is a well-known mathematical fact in learning theory and has, for instance, the consequence that there is no universal learning algorithm, as expressed by the No Free Lunch Theorem (Wolpert & Macready, 1997). According to this theorem, there is no universal learning method since the average performance of two learning methods will be the same across all possible problem instances drawn from a uniform probability distribution. If the opposite were the case, namely that there would exist a universal learner that can learn any problem by statistical reasoning, then the algorithm would be completely unbiased. But finding such an algorithm is impossible and therefore prior knowledge about the problem at hand is always required. Intuitively speaking, since all learning can in principle be expressed in terms of Bayesian inference, which requires some prior probability distribution, learning cannot happen without some prior presuppositions.

[6] This extension can be thought to be similar to the extended mind as discussed by Clark and Chalmers (1998), if the extended mind hypothesis is interpreted as a claim about information processing. Independent of whether this hypothesis is correct, the cognising system and its extension to different states, as discussed in this chapter, can be mathematically characterised in terms of Markov blankets, as discussed by Clark (2017). Notice that the equation $C_S = C_M + E + M$ is metaphorical, because the +-operator amounts to an information theoretic connection that must be made between cogniser and experimental method. How this connection is made is, however, in most practical cases unknown.

In the case of the manifest image, the experimental device is simply zero, and the information content we impose is determined by our mental state alone.

The idea of the presented account is then summarised as follows: The individuation and identification of X_M requires a certain amount of information import I_M by the cognising system. By means of conceptual and experimental work, the cognising system reaches a new cognitive state C_S, which is necessary for the detection of new entities. These methods carry a certain additional information import ΔI. The information or prior knowledge I_S required to detect objects X_S is not the same as the information I_M required to detect X_M. Since the additional information import $\Delta I := |I_M - I_S|$ can often not be quantified and thus no apparent connection between X_M and X_S can be found, there appears to be a clash between the object structures. In Bayesian terminology, I suggest that a structural clash results from a difference in the cogniser's priors when observing different levels of reality.

Before explaining the details of why it is that priors vary between the levels of description, I present a simple example in which a structural clash exists due to differing cognitive states.

16.3 Example from Thermodynamics

To illustrate how information is imposed on a system and how this is reflected in the way we characterise levels of reality differently, I present a case from thermodynamics in which it is possible to quantify the relative amount of information the observer contributes to a system at one level of description compared to another. This provides an understanding of how different domains are connected. Note, however, that this is one of the very rare cases where such quantification of information is possible, for reasons discussed below. The example originates from the work of Edwin T. Jaynes (1989), who used it as a case study in favour of a Bayesian approach to scientific practice. The case study is quite technical, so readers unfamiliar with the concepts presented may prefer to skip to the summary paragraph of the example before continuing with Sect. 16.4.

Imagine the scientist Ludwig, who studies the behaviour of a sugar crystal in water. At time $t = 0$, the sugar components—let us call them 'particles'—are arranged according to a distribution $n(x, 0)$. As time increases, Ludwig expects the distribution to flatten and become more uniform. This change in distribution is described by $n(x, t)$, whose limit over time is the uniform distribution itself (see Fig. 16.2).

During this period of reasoning, we assume Ludwig to be in the cognitive state C_M. In this state, he observes a certain diffusion current, or flux $J(x, t)$, carrying the sugar particles from regions of high to low density. By theoretical reasoning, he assumes that the change in density relates to the flux by obeying the principle of conservation:

Fig. 16.2 At time $t = 0$, point particles lying in a two-dimensional plane are arranged in a quadratic lattice. Shown are the distributions over the plane at different points in time. As time progresses, we observe a positive flux of particles towards the outer parts of the plane

$$\frac{\partial n}{\partial t} + \text{div } J = 0 \tag{16.1}$$

In addition, Ludwig phenomenologically finds that the flux can be related to the density gradient according to Fick's law:

$$J = -D \text{ grad } n \tag{16.2}$$

where D is the diffusion coefficient, which can be measured for different fluids.

Ludwig is aware that Fick's law is not a fundamental law and that the observed flux is not a fundamental entity. He therefore wonders how Fick's law might be derived from first principles and seeks to establish a connection to the particle level and the level of phenomenal experience. In other words, Ludwig is asking how $J(x, t)$ can be conceptualised and derived from a theory about the motion of particles, which lies beyond immediate phenomenal experience.

For his analysis, Ludwig turns to the *particle level of description* and considers a single particle at position $x(t)$. He approaches the problem as follows: in order to derive the flux, one can compute the average velocity of a single particle and then determine the average velocity over all particles to identify a drift. The motion of individual particles is characterised by rapid fluctuations. One can define $y := x(t + \tau)$ to be the particle's position one time-step τ after t, and $z := x(t - \tau)$ as the position a time-step earlier. If one tracks this particle over two sufficiently long time-steps, one can turn to classical mechanics and estimate the average velocity of this single particle as

$$\overline{v} = \frac{x(t + \tau) - x(t - \tau)}{2\tau} \tag{16.3}$$

The particle will move to $y := x(t + \tau)$ after the time-step τ with a probability $P(y \mid x, \tau)$. According to the Central Limit Theorem, this probability distribution can be approximated as a Gaussian distribution, since the number of steps the particle undergoes is very large. Thus,

$$P(y \mid x, I) = A \exp\left(-\frac{(y-x)^2}{2\sigma^2(\tau)}\right) \tag{16.4}$$

where A is a constant, $\sigma^2 = (\delta x)^2$ is the squared expected displacement, and I represents all prior information stated or implied by the formulation of the problem.

This model implies a mean displacement of zero, suggesting that the mean velocity of all the particles is also zero. By approximation, the model assumes that there is no interaction occurring between the sugar particles, and that it is impossible for any sugar particle to know where the other sugar particles are positioned.[7] Given the particle's fluctuating behaviour, it is then a reasonable assumption to say that its motion is symmetric around its starting position, as characterised by (16.4). Performing a computer simulation of this particle model verifies that it generates the phenomenal experiences we have, as shown in Fig. 16.2. Thus, regardless of whether we consider this model of particle motion successful, the assumption of particles moving randomly and symmetrically around a certain mean leads to a phenomenal experience of there being a flux.

Ludwig notices that there seems to be an inconsistency between the observed phenomenon and our theoretical model. Since $P(y \mid x, I)$ is symmetric, the particle is equally likely to move to the right as to the left. Consequently, $\overline{y} = \overline{z} = x$. The estimated average velocity $\overline{v}$ thus turns out to be zero, meaning that the equations of motion suggest the particle is, on average, not moving at all—no flux and no directional gradient. Ludwig is highly puzzled, as there seems to be a clear clash between the two levels of reality. Based on phenomenal experience, the assumption about symmetry appears to be clearly false, and yet it seems like a reasonable assumption at the level of individual particles. Moreover, it generates the correct phenomenal experiences when simulated by a computer. So what went wrong in Ludwig's analysis?

The solution to the problem—as implied by Jaynes—is that Ludwig finds an inconsistency because he is attempting to bridge the gap between the two levels of description directly via the connection P_X, without considering the differences in the cogniser's state. To avoid the clash and connect the phenomenal law from first principles, Ludwig must turn to the subject himself and quantify the difference in information that the subject imposes onto the system in the two cases. As Jaynes writes:

> The question is not "How do the equations of motion require the particles to move about on the average?" ... The question is: "What is the best estimate we can make about how the particles are in fact moving in the present instance, based on all the information we have?"

[7] Note that the example does not make any distinction between sugar particles and water particles besides their label. Anyone who finds this concerning can instead replace 'sugar particles in water' with 'water particles with a label A in water particles with label B'.

> The equations of motion are symmetric in past and future; but our information about the particles is not. (1989, p. 5)

The single-particle analysis is correct when we only consider the trajectory of that one particle. But Eq. (16.4) expresses the average motion over all possible motion-paths that are compatible with the dynamics. When we observe a flux, we are not simply following an individual particle but, as Jaynes explains:

> Our prior information selects out of the class of all possibilities in (16.4) a smaller class in which our particle is likely to be, in which movements to the right and the left do not have equal weights. It is not the dynamics, but the prior information, that breaks the symmetry and leads us to predict a non-zero flux. (1989, p. 5)

This is best shown in mathematical terms.

Using Bayes' theorem, we can compute the inverse probability

$$P\,(z \mid x, t, I) = AP\,(z \mid I)\,P(x \mid z, I), \tag{16.5}$$

where $P\,(x \mid z)$ is simply the same function as in Eq. (16.4). Since $P(z \mid I)$ must be proportional to $n(z)$, it follows together with Eq. (16.3) that

$$\log P\,(z \mid x, I) = \log n(z) - \frac{(z-x)^2}{2\sigma^2(\tau)} + \tilde{A}. \tag{16.6}$$

By differentiating both sides of the equation, we find that $\overline{z} \neq x$, but rather

$$\overline{z} = x + \sigma^2 \,\text{grad}\,(\log n) = x + \frac{(\delta x)^2}{2\tau}\text{grad}(\log n). \tag{16.7}$$

This equation can be substituted into Eq. (16.3), from which we get

$$\overline{v} = -\frac{(\delta x)^2}{2\tau}\text{grad}(\log n), \tag{16.8}$$

and a diffusion flux over the time interval 2τ as

$$J(x,t) = n\,\overline{v} = -\frac{(\delta x)^2}{2\tau}\text{grad}\,n. \tag{16.9}$$

It is thus thanks to Bayes' theorem and the inclusion of prior information that we can derive the correct phenomenological form of diffusion from the motion of individual particles.

Let me summarise the example. When we track sugar particles in water (or ink particles in water, or even theoretically pre-labelled water particles in water), the totality of the particles' motion is characterised by a flux. However, when we analyse the individual particle level, no flux seems to exist, since individual particles move around their initial starting point in a random but symmetric fashion.[8] Thus, there appears to be a structural clash between the two levels of description regarding the dynamic relationships between the objects in the system. Our analysis shows that the reason for this clash is that the observing scientist, Ludwig, is at different times in different cognitive states, C_S and C_M, when he is describing the phenomenon at the different levels. The relationship between the two states is characterised by Eq. (16.5), which expresses the information the cogniser puts into the system when considering not only an individual particle but a whole group of particles. The term $P(z \mid I)$ thereby contains information about which particles belong to the class of interest (sugar particles) and which do not (water particles). By making a clear-cut separation between particles and including this information in the system, new features and relations are found, such as the flux. Note again that the division of the particles into two classes is ultimately a subjective choice and not based on the physical properties distinguishing sugar and water. This is apparent from the fact that all derivations are independent of their physical constitution and also hold for pre-labelled water particles in water. To observe the flux, it is only relevant that the observer preselects some particles he or she is interested in.

16.4 Quantifying Information Differences

In the example from thermodynamics presented above, we saw that different mental priors about a physical system result in different observed structural properties. While we were able to explain and thus resolve the clash in one instance, the question now is whether the structural clash between images can *always* be explained and resolved by quantifying the differences in the observer's prior knowledge. There is both a theoretical and a practical response to this question. Let us begin with the latter.

In the example from thermodynamics, the reason why we have Ludwig being in either state C_S or state C_M is that the two forms of description required additional information, either imposed conceptually or by an empirical device.[9] In the presented case, the information difference between the two states was easily quantified because of the small difference in information between the two forms of conceptualisation. This also reflects the minor relevant contribution in information made by the experimental device needed to detect individual particles. It turned out that we only had to quantify the information import that occurs when mentally

[8] The images in Fig. 16.2 were, for instance, generated by letting particles make one step in a random direction after every time-step. This includes the possibility of a particle simply moving back and forth at one place.

[9] In this case, Ludwig did not perform an empirical investigation to determine whether ink and water can be represented as particles. However, the information was already imposed earlier by other scientists who conducted such experiments. This information carries itself into Ludwig's conceptualisation process.

selecting a set of particles rather than considering each particle individually. Thus, it was easy to compute the information *difference* between the cognitive states, and we never had to quantify the entire prior information of the two systems. Quantifying the total information of a system that includes a human observer is, practically speaking, impossible, but differences in information may still be tractable. However, the structural clash between the objects in other scientific enterprises is, in practice, much more difficult to reconcile, as it is not clear what the information difference amounts to. We have no general method of quantifying the amount of information one device, such as a particle accelerator, imposes on the system of observation in comparison with other devices such as a microscope or the naked human eye. Converting physical structure into information is, in general, extremely difficult if not practically impossible.[10] Without making progress on these questions, it is doubtful that we can make headway in quantifying the information import of scientific practice and the apparent clash between images.

I believe that this practical problem alone is already sufficient to justify the account presented in this chapter. However, one can also identify an additional, more theoretical reason why priors can usually not be brought into connection with one another.

We have seen that one concern one can have with the linguistic contextualist is that it seems unclear why there could not be a single linguistic context in which one can formulate one's ontology. Likewise, there is the question for cognitive contextualism whether the priors of our conceptualisation and our experimental devices do not simply contain different parts of the information about the world that are ultimately all compatible with one another and could, at least in principle, be connected into one unified theory. Although there clearly is a practical problem in quantifying prior information, it seems intuitive that ultimately—apart from this practical difficulty—one could arrive at a unifying theory that bridges the different levels of reality.[11] Given that the world is one unified whole and that experimental devices function reliably, it seems intuitive that experimental devices encode different prior information that is, however, always consistent with the priors of other devices. This would then imply that, at least theoretically, there is no structural clash between the different levels of description. However, according to statistical learning theory, this is not the case. As we have seen in earlier chapters discussing the findings of the statistician Charles Stein, unbiased models are often not optimal for a given learning problem.

What holds for our cognitive system also holds for the joint system of the scientist together with his or her experimental device. The prior information an experimental device puts into the system might well enhance the chances of detecting certain

[10] This is comparable to one of the great challenges currently faced in machine learning research. It remains unclear how much information different neural network structures impose on the system they are trying to learn. Certain simpler learning methods have, however, been successfully interpreted in information-theoretic terms (see, e.g., Buhmann, 2010).

[11] For example, by introducing bridge laws similar to those suggested by Nagel (1961).

patterns in the data but may also introduce bias. There is no reason to believe that this bias should be the same for different experimental devices, and therefore the structures detected through their use might well be formed on the basis of priors that differ in their bias. If this is the case, then the information encoded in different cognising systems might simply be incompatible because of differences in bias that are unavoidable without trading them for higher variance.

16.5 The Clash Between Images

With this in mind, I explicitly respond to the challenges resulting from Sellars' dichotomy. First, why are the manifest and the scientific images of the world incompatible with each other? The patterns we discover in our ordinary lives or as scientists are always shaped by the scientific methods and tools we employ, as well as by our modes of conceptualisation and prior experiences. Thus, the observer's cognitive state changes in information-theoretic terms, resulting in different patterns with different properties at different times. If the individuation of an entity, and thereby its existence, is conditioned by the cognitive state of the observer, and if this state changes over time, there is no reason to assume that such entities must share the same or even compatible properties. The problem of reconciling different layers of reality, such as the manifest image and, for instance, the world of molecular biology, is therefore not primarily a matter of matching the postulated entities and their properties with one another. Rather, it concerns the challenge of relating one cognitive state of the observer at a given time to another.

Second, given that scientific work always begins within the manifest image, does this image deserve a special status? The manifest image is merely one representation of the world, corresponding to a particular cognitive state. Due to our biological constitution, the manifest image commands special attention, since it is the world we encounter without the use of additional instruments, apart from occasional tools such as eyeglasses or hearing aids. It is important to note, however, that the objects we find within the manifest image can vary significantly in their structure depending on societal customs and the kinds of actions performed within a society. For example, the fact that we typically individuate a hammer as a single object, in contrast to a pen and its lid which we might instead regard as two distinct objects, is a result of how we use these items.[12]

Third, what is the ontological basis for scale in a scale-relative ontology? The existence of the patterns a scientist discovers in the data depends on the cognitive state of the investigator and thus also on the methods used to detect those patterns. Since instruments vary in precision and sensitivity, it is not surprising that different sciences produce theories that are successful on specific measurement scales. Nor

[12] Quine (1960) and Putnam (1981) famously argue that the structure of material objects is determined by language use and social practices.

is it surprising that even within a single scale, a science may be highly non-unified when different instruments are involved.[13] Consequently, if no level of description is to be privileged while others are dismissed, as is the case with the conservative naturalist, then the ontological basis for the entities discovered in the empirical sciences must be treated as relative to the observer.

Summary

The different levels of reality are structurally irreconcilable, which led Sellars and the linguistic contextualist to draw clear distinctions between them. However, this approach struggles to account for the fact that macroscopic occurrences cannot be treated as independent from physical ones. The levels of reality are clearly connected and interdependent. For the contextualist to offer a convincing account, they must explain the details of this connection.

In this chapter, I have argued that the levels of reality result from different experimental and cognitive processes involved in empirical research. These processes alter the observer's cognitive state and thereby affect the detection of patterns. The structural inconsistencies between different levels of reality are thus the result of differences in the prior knowledge or information the observer contributes to the system. Once we include the observer in our analysis, we find that the inconsistencies dissolve once we quantify the differences in information applied to the system under investigation by different observing systems. In contrast to the liberal naturalist who denies the existence of a clash, the view presented here accepts that a structural clash exists, yet avoids the need to assume a fundamentally disunified world.

[13] The fragmented nature of the sciences has been well documented. For instance, the disunity of physics is one of the central themes in Nancy Cartwright's (1983) seminal work.

Chapter 17
Reality and Logic at Conflict

In this chapter I deepen the discussion on the connections between idealism and the puzzles about ordinary objects that have been investigated so far. I will first discuss several issues about the motivation for idealism and what response one can give to a variety of realist objections that have not yet been covered. I will then elaborate on how we ought to think about the puzzles we find under discussion in contemporary metaphysics and the logic literature.

17.1 The Essential Qualities of Ordinary Objects as Constructs

As mentioned earlier, the retinal stimulation is underdetermined by its cause. Thus, there are a multiplicity of possibilities of what the experienced qualities of ordinary objects could be. We saw that this is illustrated by a variety of different clinical cases in which patients in fact see objects differently despite having complete access to the visual data.

Although there are many possible ways to experience the world, given the data we receive through our senses, most of us see the world to consist of ordinary objects that have certain essential qualities. The reason why we see ordinary objects to have the qualities here under discussion and why we find them essential is *not* because these qualities are extracted from the data by a certain innate feature detection machinery that makes the right presuppositions on the data due to evolutionary reasons. Instead, the modern brain sciences imply that these essential qualities can be explained to be a consequence of innate RPPs. Given that there are multiple different possibilities of how our experiences and thus our object concept could be formed, and given that many essential qualities of ordinary objects can be derived from the perceptual processes alone, it would be a major coincidence if these qualities we ascribe to ordinary objects were objectively true. This debunks

A. C. Schwaninger, *Cognitive Metaphysics*, Synthese Library 529,
https://doi.org/10.1007/978-3-032-05420-3_17

the common-sense assumption that mind-independent objects have properties like having spatio-temporal persistence, being not collocated with each other, and so on.

The metaphysical realist will object to this argument that even if we agree that the brain does not track pre-given features in the world but rather constructs the features for which it is sensitive, the construction process itself can be interpreted in a way that we arrive at a realist understanding of cognition. In other words, the objection might be that even if there are no pre-given features that our brains are tracking, the RPPs *themselves* are the result of evolutionary pressures such that we can accurately see the features of ordinary objects within our ecological niche.

Besides the fact that evolutionary pressures are not driving representations towards truth, and besides the fact that representational biases are often more successful than unbiased representations, this argument misses the point that it is not due to evolutionary selection processes that RPPs are essential to cognition. RPPs are not formed because of any specific evolutionary pressures. Instead, RPPs enable evolutionary selection processes to take place in the first place, because RPPs are the preconditions for life in general. They are not the result of natural selection but one of the presuppositions for natural selection to be possible in the first place. According to the FEP, thermodynamics and information theory dictate that living systems operate by predicting processes to minimise variational free energy, independent of the course that evolution might take. It is for this reason that RPPs are universal and exist in all cognitive systems, independent of any specific evolutionary pressures within the specific ecological niche. It can therefore not be the case that the structures resulting from RPPs are a consequence of specific evolutionary pressures of an environment with certain properties. Thus, RPPs and the resulting experienced structures would only be coincidentally in line with some *true* mind-independent structure in the world. Facts about RPPs are derived from the FEP, which only relies on facts about thermodynamics, information theory, and conceptual work about living beings. Facts about the essential properties of RPPs are not derived from ordinary objects, because the FEP does not make any assumptions about ordinary objects in the first place, and the essential properties of RPPs cannot be the consequence of any evolutionary pressures within our ecological niche. Therefore, the category of RPPs neither depends on ordinary objects nor on evolutionary pressures that could determine the properties of RPPs. This undermines the realist's objection.

17.2 The Universal Status of Predictions

The results that were derived about the nature of ordinary objects are universal amongst living systems at a certain level of complexity. From a methodological perspective, Cognitive Metaphysics focuses on the cognition of *Homo sapiens*. But Kant's aspirations are to find a theory of greatest generality. For Kant, the features of the objects of our experience are *necessary* in the sense of being essential to all objects of our cognition, not in the sense of being contingent to our specific biological faculties. According to Kant, his metaphysical principles hold for any objects that finite beings can experience.

Although PP concerns the cognitive processes of humans and those animals that possess a similar neural physiology, the FEP makes no assumptions about the cogniser's physiology. According to the FEP, *all* living beings in the universe are systems that must minimise free energy in order to sustain themselves (Friston, 2013a). It is argued on the basis of a transcendental argument (as formulated in detail in Colombo & Wright, 2018), which draws conclusions about computational processing from the fact that living systems exist. This has the mathematical consequence that any living being at least unconsciously aspires to make predictions about events in its ecological niche. Since this is a theory about all living systems, it cannot be empirically verified by observing a specific living thing but rather follows from mathematical analysis, the fact that entropy always increases over time, and the conceptual truism that living beings are sustaining systems.

Consequently, it is not the case that the category of RPPs is limited to humans or human-like animals. Instead, all living systems that are capable of having experience and knowledge do so by making predictions. This is not a logical necessity, nor a necessity for all cognisers in all possible worlds. But it is a necessity for any living being in our known universe, which, for all we know, is governed by the laws of thermodynamics as well as the laws of information theory. Thus, this result carries metaphysical consequences within the Kantian framework that is concerned with the actual universe because it sets the preconditions for our knowledge of objects.[1] The FEP characterises the cognitive process shared among all living beings with cognitive abilities, and with this, the category of predicting processes deserves transcendental status.

Does it therefore follow that all living beings, including ants and trees, cognise ordinary objects the way we do? With regard to the essential structures we discussed earlier, the answer is 'yes', if predictions are processed optimally in a robust way. But this is surely not the case for systems that have not evolved to make such perfect predictions; thus, the answer is, generally speaking, 'no'. There are different possibilities for how predictions can occur in nature, and surely not all predicting processes are *robust*. The assumption in my analysis has always been that we presuppose a cognising system which makes predictions that are robust to noise. In simple organisms, however, predictions about the environment are merely made through the genome, and the organism has no ability to alter its feature detection machinery. Consequently, the objects of perception can have different properties, such as having sharp, detectable cut-offs in a sorites sequence.

The ability to make robust predictions is ultimately a matter of degree. Some organisms process information in a way that shares the human ability to generate

[1] Note that within the FEP framework, living beings are not treated as ordinary objects but rather abstractly interpreted as a set of nodes within a so-called 'Markov blanket'. While the computational processes of these mathematical objects rely on some physical manifestation that can be an ordinary object like a robot or an animal, the FEP does not characterise these mathematical objects as having the properties of ordinary objects that are here under discussion.

predictions that are highly robust, whereas other organisms process information by making predictions in a far less robust way. Whether the human brain has found the universally optimal method to minimise average prediction error in a robust way is, of course, difficult to say. But at least among all living beings that share the same degree of robustness when making predictions, the experience and the related concept of the essential qualities of ordinary objects are shared.[2]

17.3 Further Objections Against Idealism

A second objection might result from the following questions: What are neural processes predicting if not ordinary objects? Does not the notion of *prediction* already assume that we are predicting things in the world, and must we therefore not be realists about these things? This objection misses the point that what neural processes are predicting is the activity of other neurones, and at the lowest cortical region, the sensory data. I never said in my discussion that I was an anti-realist about the sensory data falling onto the retina (something that even a person suffering from visual agnosia can see). I am only an anti-realist about the essential structural qualities that are ultimately not found in the sensory data itself but are essential to our ordinary sense-making of the world and have been important in metaphysical debates. I am thus not claiming that *all* properties of the world or sensory data are to be conceived from an anti-realist perspective.

As a third objection, the realist could suggest that merely because the essential qualities are derivable from RPPs does not logically imply that these essential qualities do not also belong to the objects of the world itself. Although this is true, the argument offers no reason why one should believe that objects in the world might ontologically possess the qualities under discussion. As suggested above, there are a multitude of different ways we could be experiencing ordinary objects. It would therefore be a major coincidence if the objects we perceive had identical properties to the objects that exist independent of us and would require an unreasonable leap of faith.

Finally, the metaphysical realist might stress that the account of perception laid out here might be valid for a large number of our perceptions but not necessarily for *all* perception. Thus, some of the objects we perceive might still be the objects as they are in the world, in the same way as our perception is sometimes misguided. But this view is mistaken. One of many reasons why we assume that the brain must *constantly* form predictions based on categories it imposes onto the data is because it explains how the brain can deal with noise. Every time we walk around or move

[2] Note that according to Burge (2014), living beings can be said to have a mind if they exhibit perceptual constancies. If this is true, then the mind starts where predicting processes are sufficiently robust. In that case, the qualities of ordinary objects which we find essential are essential to all beings that have a mind.

our eyes, there is a 'dramatic change of sensation' (Frith, 2007, p. 201), and yet the world appears to be stable. This stability is ensured by the brain predicting the retina's sensations by taking the subject's own always-present movements into account.

This final argument can be turned around against realism: Imagine it were the case that the brain would allow us to see the world as it is. We could clearly not expect that perception would always operate accurately, and thus, sometimes errors, as in the case of optical illusions, must be expected. However, if the brain were to always try to offer us an image of the world as it is, any such errors that were to occur would not be expected to reflect on the essential properties of how we see the world. If it were the case that our perceptual system is presenting us with how the world is, then there is no reason why erroneous cases in perception should contain any structures that could be traced back to properties of the brain. We would rather expect to see noisy images or a colourful error screen, as we find them on television sets. But what we really find is exactly the opposite. Consider Fig. 17.1, in which one of the categories that the brain imposes conflicts with another category because both categories have a high probability of leading to good predictions. Consequently, the brain is undecided and alternates between two possible ways in which composition occurs.

The fact that we see the presented images in such peculiar ways provides evidence that PP is correct for all perception and is *constantly* imposing the qualities we ordinarily ascribe to objects in the world. PP was not developed to explain such anomalies; rather, these anomalies are predictable consequences of the theory. If such cases did not exist, then we would have reason to doubt the validity of PP. Conversely, if we did see the world *as it is*, we would need an explanation for why visual anomalies arise—anomalies in which compositionality occurs precisely when it aligns with the subjectivist or anti-realist interpretation I am offering here. It is therefore up to the realist to explain how such anomalies come about, and why the

Fig. 17.1 Image depicts a part of Salvador Dalí's painting 'Slave Market with the Disappearing Bust of Voltaire' (1940)

processes that reliably generate them should not be regarded as constitutive of our perceptual experience more generally.[3]

17.4 The Cuts of Thought and Language

In the first part of this book, I introduced a variety of puzzles about ordinary objects. The solution I offer to the various paradoxes concerning material objects is a solution in the sense of offering an explanation of how it is that we see the world as we do—how problems of identity, the sorites paradox, the problem of the many, and so on, arise in the first place, given how the mind works. I do not think there is much more we can reasonably wish for regarding these problems. In his thousand-page investigation of various logical paradoxes, Ulrich Blau (2008) arrives at the conclusion that some logical paradoxes are not convincingly solvable within a logical framework and I agree that many problems must be examined from a psychological perspective rather than a purely logical one. In his explorations of vagueness, he explains that the most basic logical framework is binary logic L. Upon further reflection, when one encounters vagueness and categorical mistakes, one transitions to three-valued logic LN, which contains, in addition to L, a neutral category. But by introducing a new category, the problem of vagueness does not disappear, because transitions to the neutral element are again fluid: 'the fluid transition becomes with every refinement a *contradiction in adjecto*. This insoluble contradiction is the common theme of Zeno's paradoxes of movement and the paradoxes of vagueness' (p. 192). While we try to make sense of ordinary reasoning by introducing a logical apparatus, the dynamics of language are context-dependent, and our subjective willingness to alter context is not always logically analysable. Ulrich Blau writes that when we reflect on material objects, we move along the following path of reflection (p. 194):

Reality → Perception → Language → Informal logic → ... → LN → L

With each step along this path, we distance ourselves further from reality. Each step 'generates cutting separations and solidifications that are justified by nothing besides their inevitability' (p. 195). Through the process of reflection, the mind draws cuts where none are to be found in reality. Pure logic cannot help us make sense of these distinctions. Yet, in cases such as the sorites paradox, probability theory—

[3] The cases presented are similar to what, since Wittgenstein's *Investigations* (1953), has been referred to as 'aspect perception'. It has been argued that aspect perception exemplifies the role we play in the constitution of the perceived world (Baz, 2020). One might object that the possibility of shifts in perceptual experience does not generalise to all perception but is confined to isolated cases. Possibly, such cases could be dismissed as mere 'illusions'. However, as some have correctly suggested, aspect perception is a feature of seeing that is always operative: 'the striking case of the change of aspects merely dramatises for us a feature (namely seeing as) which is present in perception in general' (Strawson, 1974).

understood as the foundation of a process-oriented ontology in which events mark moments of surprise when models are updated—can provide the necessary tools.[4]

While Blau is only able to provide a negative result from his profound research investigation, the answers I provide are a positive response to the paradoxes we face about ordinary objects by explaining how these puzzles come about. My results are closely aligned with Blau's views, but using modern theories of cognition, I was able to show how our intuitions and thoughts about ordinary objects arise when we are confronted with raw sensory data. I have shown that it is thanks to our cognitive architecture that we impose certain structures onto the data, and that these structures do not exist mind-independently and are in this sense non-veridical.

A person suffering from visual agnosia is someone who does not see these structures. The reflection process characterised by Blau ends abruptly at the very beginning, before a completely structured perceptual image is formed. In this sense, patients suffering from visual agnosia are 'closer' to seeing the world as it is. Their reflection processes do not allow them to reason about the world, but they can see the world more accurately than ordinary people, at the cost of not being able to function better than a blind person.

Summary

The debunking arguments in favour of idealism can defend themselves against a variety of realist concerns and follow the general structure of a transcendental argument. This chapter shows that the puzzles concerning ordinary objects are not always solved in a strict logical sense, but by providing an explanation of how they come about through the subject's flexible mechanisms of carving up reality in ways that make it more predictable.

[4] An obvious objection is that probability theory depends on logic and therefore, if it can solve anything about ordinary objects, so must logic. But I find the successes of logic in this domain underwhelming, and would rather raise the question: might probability theory be more fundamental than logic? In an article by Fields Medalist David Mumford (2000), he offers compelling reasons to believe that probability should be placed within the axiomatic foundations of mathematics, rather than treated as something that rests on set and measure theory.

Chapter 18
Philosophising and the Clash

Philosophers can engage with the question of what it means to see and perceive without ever mentioning the cortical activity involved. They engage with the question of what it means to perform an action without any interest in the neural underpinnings that result in bodily motion. They engage with the question of what it means to have knowledge without ever bothering about the precise belief formation processes involved. As we have seen throughout this book, many eliminativists tend to dismiss this type of work. While I have so far provided an account of how to understand the clash in the material world from a neuroscience and machine learning perspective, the question I raise in this chapter is how we ought to think about this conceptual clash in philosophy. What I argue is that the FEP and PP even offer us an understanding of what most philosophical engagement amounts to, and explain in what sense this activity should not be dismissed, as the eliminativist might suggest.

In an earlier chapter, we have seen that intuitions are problematic in metaphysical work and that the excessive use of intuitions in metaphysics challenges the relevance of the discipline. This is because metaphysics, in the sense I have presented it at the outset of this book, is not purely conceptual but engages in developing models that ought to be somewhat connected to the world we live in. In contrast, most other parts of philosophy place a greater emphasis on conceptual work, which only appears to the eliminativist to be problematic. The reason for this is that when we develop concepts, intuitions are a source of evidence on which we can rely. How can we understand this more precisely?

18.1 Intuitions in Philosophy

In everyday life, many of the decisions and actions we take happen unconsciously, and many of our judgements rely on what we call intuitions. This term is used in a variety of different contexts: we might find the handling of a new mobile

A. C. Schwaninger, *Cognitive Metaphysics*, Synthese Library 529,
https://doi.org/10.1007/978-3-032-05420-3_18

device intuitive. A brilliant chess player might have great intuition for finding the best move. Or we might intuitively connect better with people who are similar in character.[1] When formulating arguments in our daily lives as well as in academic practice, intuitions are often employed as preliminary assumptions.

In contrast to what we might expect from a discipline that boasts about having an unprecedented aspiration for precision, many philosophers would even agree that the appeal to intuitions in the analytic tradition is omnipresent and that its significance can hardly be overstated:[2]

> George Bealer does it. Roderick Chisholm does it a lot. Most philosophers do it openly and unapologetically, and the rest arguably do it too, although some of them would deny it. What they all do is appeal to intuitions in constructing, shaping, and refining their philosophical views. (Kornblith, 1998, p. 129)

We can turn to any domain in analytical philosophy and find assumptions in arguments that are made without further explanation or reference, since they appear to be self-evident. Such statements might be 'if P then not not P', 'torturing children for fun is wrong', or 'there are no five-sided squares' (Pust, 2012). The reliance on using or forming intuitions becomes even more apparent when looking closely at the thought experiments philosophers develop, which carry their force by appealing to certain intuitions rather than by providing rigorous arguments.

A straightforward example is the classical Gettier cases, which show in the form of thought experiments that there is more to knowledge than having a justified, true belief. One such thought experiment runs as follows:[3] Imagine you walk past a clock at 9:00 and read off the time. The clock says that it is 9:00, and therefore you believe that it is 9:00. You are clearly justified in your belief because you are reading the time from a clock that is generally reliable. By coincidence, however, the clock is not functioning that day. In fact, it stopped functioning the day before, precisely at 9:00. It is thus only coincidental that you are right in your belief that it is 9:00, because that is the time at which you passed the clock and decided to read off the time. According to most people's intuition, it would be wrong to say that you are actually knowledgeable about the time despite having a justified, true belief. Therefore, knowledge must amount to more than having justified, true belief. The argument can be summarised as follows:

---

[1] For a detailed analysis of our everyday use of 'intuition' and a contrast to the use in philosophical literature, see Cappelen (2012).

[2] Some philosophers are an exception to this and believe that the role of intuition is limited or insignificant to the discipline. See, e.g., Cappelen (2012), Williamson (2007), and Deutsch (2009, 2010).

[3] Note that this is Bertrand Russell's (1948) stopped-clock thought experiment and not part of the famous Gettier (1963) article. It is, however, simpler to understand and of the same kind as Gettier's original cases.

P_1 : You have a justified, true belief that it is 9:00.
P_2 : You do not know that it is 9:00.
C: Knowledge is not justified, true belief.

The strength of P2 is based solely on intuitive grounds, and our intuition that P2 is true is typically very strong even when we have no concrete explanation for why this is the case.[4] Other similarly straightforward examples of thought experiments include the intuition of saving the child in Singer's shallow pond analogy, the intuition of attributing new knowledge to Mary the colour scientist when she leaves her room, or the intuition of refusing to attribute intelligence to Searle's Chinese room.

18.2 Discoveries by Intuitions

Given that philosophical theorising heavily relies on our intuitions, many philosophers have questioned the extent to which they can be taken as a source of *evidence*. What is at stake appears to be the legitimacy of a key ingredient in much of philosophical theorising, and if it were to turn out that intuitions cannot be trusted, then large parts of the philosophical enterprise would prove mistaken.

At least for Alvin Goldman (2007), it was clearly a 'landmark discovery' (p. 1) in the history of philosophy when Gettier showed that knowledge is not equivalent to justified, true belief. The discovery did not arise merely from the formulation of thought experiments, but largely from the fact that almost everybody reading Gettier's examples shared the same intuitions. Had it been otherwise, such that people's intuitions had not converged, then there would have been no discovery. Thus, given that philosophical intuitions are occasionally reliable indicators of facts about our concepts, they can be treated as a source of evidence (Goldman & Prust, 1998, p. 189).

Gettier's discovery is not a scientific one but a discovery coming from philosophical analysis. But what is such a discovery about? In antiquity, it might have been fashionable to think that the target concern of philosophical inquiry lies in mind-independent facts about a non-physical world. For instance, a Platonist might have imagined that Gettier's discovery concerned forms existing in some non-material realm. However, since Plato's heaven does not exist in space and time, such a

[4] Notice that studies have been conducted which suggest that the intuition supporting P2 is not universally shared. While a majority of undergraduate students at Rutgers University with Western backgrounds judged P2 as true, a majority of East Asian students at the same university drew the opposite conclusion (Weinberg et al., 2001). However, whether these experiments actually show that different people differ in their understanding of the concept of knowledge, as suggested by the experimentalists, can be disputed. It might well be that they simply did not properly understand the described situation during the experiment.

proposal seems doubtful, as it is unclear how we could have epistemic access to such entities.[5] Instead of gaining insights into knowledge *itself*, Gettier offers us insights into our *concept* of knowledge. Intuitions in philosophical analysis can thus help us investigate concepts understood as things possessed by individuals of the speaker-community and inseparable from the human mind.[6] This does not imply that these concepts are *purely* subjective entities. They clearly have a dimension that transcends the individual subject due to their normative force; their application can be correct or incorrect. Nevertheless, they must be anchored in some individuals' brain states. Even if a concept is defined by a speaker-community, its meaning is determined by the subjects of this community interacting with one another and thereby reinforcing the concept's meaning through their usage and possession. Gettier's intuitions have therefore led to a discovery about a concept in a psychological sense.

18.3 The Nature of Conceptual Intuitions

Such discoveries concerning the limits of the applicability of our ordinary concepts are where philosophical analysis shines the most. It is this domain to which no other discipline is dedicated, and no other research method appears to offer comparable results.[7] It is for this reason that much of contemporary philosophical thought about intuitions concerns a very narrow sense of intuitions, in which intuitions are conceived as having a special phenomenology that possesses epistemic value.

Let us call intuitions about the use and applicability of concepts *conceptual intuitions*. These can be characterised at the experiential level as follows. We as subjects have a conceptual intuition about the use and application of a concept when we do not require any further conscious inference to justify it to ourselves. The fact that we find a proposition about concept application so compelling appears to come from nowhere, as the reasons are opaque to introspection, yet we somehow feel a certain pressure to accept the proposition. This pressure does not force us to believe in the proposition we find so seductive; however, it creates the disposition to do so.

Conceptual intuitions are by no means unique to philosophical inquiry. In fact, we experience them all the time. Consider the first two verses of this poem by William Blake, in which the last word is left blank:

---

[5] It also seems unlikely that Gettier made a discovery about knowledge understood as an abstract entity existing independently of the human mind in some other sense of objectivity. For a detailed discussion on objectivism about concepts, see Glock (2010, pp. 312–315).

[6] Many would argue that what I am saying applies to all concepts. However, for the present purposes, I do not want to be over-committed about what concepts are in general.

[7] Note that experimental philosophy might extend the classical methodology by informing us about ordinary judgements among non-philosophers, but it certainly does not replace the classical methodology.

Tyger Tyger, burning bright,
In the forests of the _______.

Even if we have never read these lines before, we have a strong intuition that

Tyger Tyger, burning bright,
In the forests of the **day**.

is wrong because it does not match our expectations regarding rhythm. Likewise,

Tyger Tyger, burning bright,
In the forests of the **right**.

is wrong because the inserted word makes no sense in context. The blanked-out word is clearly 'night' numerous reasons for our having the right intuition about this. First, we feel that the sentence's grammatical structure requires us to insert a noun rather than, for instance, a preposition. Second, the semantic context makes it far more likely that the blanked-out word must be 'night' rather than 'dessert', 'brain', 'day', 'Bandipur National Park', or 'highland region'. Third, the repetition of the word 'Tyger', as well as the metrical pattern, creates a rhythm that makes us anticipate and feel the need to insert a monosyllable. These different cues allow most people to have the same conceptual intuition about which word best completes the blank space.

Whenever we utter words and sentences or must apply a term to a given situation—such as in a Gettier case—our cognitive system essentially does what we consciously do when reading Blake's verses and trying to fill in the blank space. It anticipates the word by using the available cues and making probabilistic inferences about which term applies best. In the Gettier cases, the subject is presented with a certain context, and the cognitive system does not judge the word 'knowledge' to be adequately applied to the given situation. Accordingly, we should not be astonished that recent neuroscience research indicates that language use and comprehension amount to an unconscious process of making good probabilistic predictions about what the next word in a sequence, given a particular context, is likely to be: '[Computational] models that perform better at predicting the next word in a sequence also better predict brain measurements' (Schrimpf et al., 2021, p. 1), of a subject reading a text at self-pace.

An understanding of conceptual intuitions as the result of unconscious processes that compute the likelihood of words being adequately employed in a given context aligns with many philosophical accounts of conceptual intuitions. These accounts interpret conceptual intuitions as attitudes towards a certain proposition that appear to be true without the subject making any further conscious inferences, as they are taken to be the result of unconscious events. In other words, the suggestion is that when we talk about our conceptual intuitions concerning the validity of some proposition P, we are referring to our attitude (or conscious mental state) towards P, which is caused by some unconscious process of which we are unaware, and that results in a disposition towards believing that P. A conceptual intuition is thus, roughly, an attitude in which a proposition presents itself as true (Bealer, 1998, p. 207). Or, if we wish to describe the unconscious level as having a more active role in determining our thoughts and actions, we can say that when we have a conceptual intuition, we are *pushed* towards accepting a proposition without being aware of the underlying, unconscious reasons (Koksvik, 2020). As in the case of Blake's verses, we have the strong inner feeling that the blanked-out word cannot be 'day' but must be 'night', despite there being no logically necessitating reason to prefer one over the other. It is simply that certain cues generate a context in which we find one word more appropriate or better placed than another.

18.4 The Heterogeneous Nature of Intuitions

In metaphysics, intuitions play a different evidential role compared to other domains of philosophy. Metaphysics differs in its aspirations from many other branches of the discipline. As we have seen, Gettier's discovery was not about knowledge itself but about the concept of knowledge. It concerned how the concept of knowledge is correctly applied or not applied to a given situation, based on how our culture uses the term. However, not all philosophy claims to be concerned with the analysis and refinement of concepts. Metaphysicians in particular claim that they are concerned with the most fundamental structure of reality and are committed to the task of identifying the categories that correctly describe the world, rather than engaging in the conceptual analysis of ordinary terms (Sider, 2011, ch. 1). For them, intuitions about concepts add little or nothing to what their work is targeted at, namely, reality.[8]

Furthermore, intuitions used in metaphysics are not conceptual intuitions and therefore differ in kind from those employed in most other domains of philosophy. Unfortunately, the dialectic of the debate concerning the nature and epistemic value of intuitions in philosophy is somewhat misguided, as it treats intuitions as a single phenomenal state rather than referring to them as a heterogeneous set of mental phenomena. As Jennifer Nado (2014) suggests, it is implausible that the intuitions of all areas of philosophy constitute such a uniform psychological kind. In the case of metaphysics, the intuitions used are largely not intuitions about concepts.

The intuitions employed in metaphysics do not arise from our subconscious understanding of certain concepts but seem to have a different origin. It is due to our perceptual experiences that we have the intuition that chairs and tables exist, rather than any strange objects. It is also due to some conscious experiences we have that we find the A-Theory intuitively more plausible than the B-Theory. I refer to these types of intuitions as *experiential intuitions*, which can best be characterised as spontaneous, ordinary judgements. In contrast to conceptual intuitions, experiential intuitions do not directly push us towards making certain utterances.

[8] Note that much of analytical metaphysics does conduct conceptual analysis to some extent. However, it is traditionally agreed that this is not the main target of the discipline, but merely required for the purpose of having relevant concepts at one's disposal.

Fig. 18.1 Ebbinghaus illusion

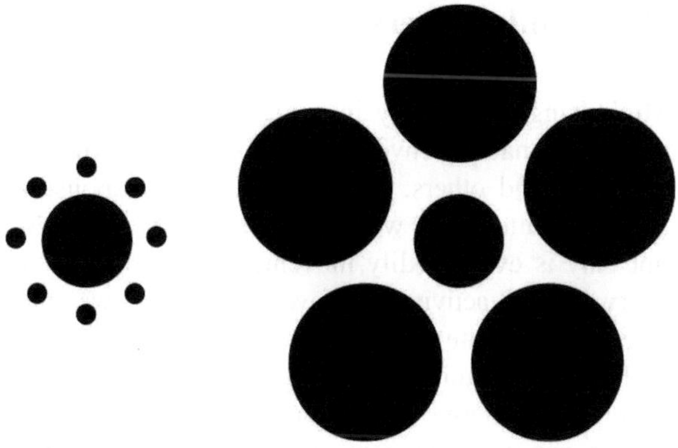

This is best illustrated by an analogy to optical illusions. When looking at the Ebbinghaus illusion in Fig. 18.1, we clearly have the experience that both central circles differ in diameter. Yet, after measuring the two diameters and confirming that they are equal, we still have the experience of the two circles being of different size, though we do not feel any pressure to make such a claim aloud. If we are presented with the sentence

'The two central circles in Fig. 18.1 have ______ diameter.'

we do not feel a strong intuition that the blanked-out phrase must be 'the same' rather than 'a different'. It is not the case that we experience a great inner pressure pushing us towards filling in the blank in any specific way. This is because our experiential intuitions are not directly connected to the words we utter or the concepts we find appropriate in a given context.

In accordance with this analogy, metaphysicians can easily distinguish between the intuitions they have and the beliefs they form based on theoretical considerations. They can have the inner feeling of time passing while believing the opposite. In contrast, conceptual intuitions tend to correlate with the beliefs we form. Thus, although there is convergence in our intuitions both on the conceptual and the experiential side, judgements and beliefs tend to converge only in the presence of converging conceptual, rather than experiential, intuitions.

Psychologically, this is explained as follows. Given the conceptual abilities of a competent speaker, a linguistic context generates an intuition about word usage and application conditions, as described above in the context of Blake's verses. These intuitions are dispositions that push us towards a certain attitude regarding the proposition we find intuitive. We trust our intuitions because they reflect the conceptual abilities we have largely acquired at a very young age. If these abilities were to change, then our intuitions would change as well, and with them also our beliefs. In the case of experiential intuitions, the transition from intuition to belief does not occur automatically because beliefs are linguistic, and our capacity to have certain experiences does not directly affect our linguistic beliefs or vice versa.

18.5 Philosophers Are Minimising Free Energy

Minimising free energy in a social context is no different from minimising it in a deserted natural environment. During both listening and speaking, we predict ourselves and others. Depending on the response and our level of surprise, we adapt the utterances we make (Friston & Frith, 2015). If the FEP is true, then not only is every bodily movement an attempt to minimise free energy, but also every mental activity and every utterance we make serves the same purpose. Therefore, despite the differences between conceptual and experiential intuitions, both forms of intuition are ultimately shaped by the underlying attempt to minimise average free energy. Whenever we have an experiential intuition, our visual system shapes our experience in such a way that we can form beliefs that are most useful for minimising free energy. Likewise, when we have a conceptual intuition, our unconscious system detects that some utterances or sequences of words are particularly right or wrong in a given situation, that is, they are expected to minimise average prediction error.

For instance, reading a Gettier-like thought experiment creates a certain context in which uttering the sentence 'You *know* that it is 9:00' does not feel right. Forming this mental thought or uttering these words is not the appropriate activity for minimising free energy, given how we humans have been socially trained to use certain terms. Put differently, we feel inclined to say that the blanked-out word in '"You know that it is 9:00" is ______' must be 'false' rather than 'true'.

The FEP connects our perceptions, like the sentences we hear and read, with our emotions and actions into one single framework. This is how it manages to explain why we find some philosophical accounts more plausible than others and what the core causal roots of our intuitions are. The activity of conceptual analysis and engineering is one that enables more coherent and thus more predictable linguistic systems, thereby helping to minimise free energy. As long as philosophical debates are not targeted at platonic forms or an ontological counterpart, there is no fundamental clash between philosophical conceptual work, which develops theories about how terms are more predictably used, and scientific theories, in which terms are more intimately related to the world.

Summary
In contrast to many other domains of philosophy, metaphysics relies on experiential rather than conceptual intuitions. Due to their nature, experiential intuitions do not provide direct evidence for particular views, and consequently, metaphysical theories are often counterintuitive. By contrast, conceptual intuitions have epistemic value in those areas of philosophy that do not aspire to ontological insights. In these domains, philosophy contributes to free energy minimisation by engineering concepts to be more predictable.

Chapter 19
Summary

In our everyday lives, when we look at the material world, we find it to be a place consisting of ordinary objects that all share structural qualities such as persistence through time and space, the ability to survive the loss of some of their parts, and the impossibility of being collocated with one another. This description is partly inconsistent, or *clashes*, with certain domains of science, most notably physics (Chap. 2). Fundamental physics describes the world as largely empty, made up of a relatively small number of particle types that propagate through space according to equations of motion. While philosophers of physics may disagree on whether these fundamental objects are individuals, there is no doubt that we ordinarily perceive the world as consisting of individuals that possess essential structural qualities.

In this book, I argue that the structural qualities found within the manifest material world are not features of a mind-independent reality but are instead shaped by subpersonal neural processes and can be analysed from a computational perspective. Drawing on recent research in neuroscience, I show that PP provides the theoretical basis for a debunking argument, dismissing ordinary objects with their structural qualities from the domain of mind-independent ontology. As an alternative, I develop a mind-dependent ontology that offers simple and unifying solutions to traditional metaphysical puzzles. It proposes an ontology of (cognitive) processes that determines the essential structure of (most) material objects.

Part I discusses the metaphysical relevance of ordinary objects and shows how metaphysics in the analytic tradition struggles to account for these entities. Many metaphysicians investigate the structural qualities of ordinary objects and frame puzzles as if these qualities were mind-independent. They debate competing frameworks for accounting for persistence or non-collocation, yet they neglect the extent to which such qualities arise from the way the brain processes sensory input, thereby contributing to our experience of objects persisting or being non-collocated. If we have reason to believe that these structural qualities are mind-dependent, then metaphysical puzzles might be resolved by investigating the neural code of the brain.

A. C. Schwaninger, *Cognitive Metaphysics*, Synthese Library 529,
https://doi.org/10.1007/978-3-032-05420-3_19

This investigation may explain why we experience the world in a certain way and clarify how such puzzles arise.

Chapters 3 and 4 introduce an eliminativist position on ordinary objects, grounded in sceptical concerns. However, as discussed in Chap. 5, adopting such a position would be a mistake. Metaphysics is not solely concerned with existence questions but must also provide an ontology that is useful for conceptualising the world as we experience it. Although radical alternatives such as eliminativism are possible, I argue in Chap. 6 that a more adequate framework should account for manifest objects and resolve traditional puzzles in a way that is consistent with our scientific understanding of the world. This leads to a form of naturalised Kantian idealism, presented under the label *Cognitive Metaphysics*. This position is idealist in the sense that the essential qualities of ordinary objects are traced to human cognition rather than to the properties of a mind-independent world (Chap. 7). However, it does not deny the existence of mind-independent matter. Rather, it holds that our experience of matter as a collection of discrete individuals with properties such as non-collocation and persistence is the result of mental processing.

According to Cognitive Metaphysics, the metaphysical nature of ordinary objects must be investigated by first identifying innate and reliable cognitive processes, second by describing these processes in algorithmic terms, and finally by deriving the essential qualities of the objects of experience from this algorithmic formulation (Chap. 8). If the qualities of the objects of experience can be derived solely from how our cognitive processes operate, rather than from the way the world is described by our best physical theories, then these qualities are said to have been debunked as mind-dependent. The strongest motivation for the proposed form of idealism comes from the debunking argument, which is initially motivated by sceptical concerns and developed in detail in Chap. 13. I show that, in light of PP, there is no good reason to believe that the structural qualities of ordinary objects are determined by how the world is, and that metaphysical reasoning about these qualities is in fact reasoning about experiences shaped by cognitive processes.

To adopt the Cognitive Metaphysics approach within contemporary metaphysics, Part II provides a detailed overview and interpretation of modern neuroscience research, focusing on PP and how this theoretical framework contrasts with traditional views on cognition. Based on early models in brain science, it is argued that classical views of cognition typically rely on three assumptions (Chap. 9). First, the brain has an innate propensity to extract the features of objects as they exist in the world. Second, it possesses a built-in feature detection mechanism that enables us to identify ordinary objects. Third, perception is characterised as a predominantly one-directional causal process. Classical views on cognition typically add a fourth assumption which is that objects are first individuated and then categorised based on their similarity to previously perceived objects. Although these assumptions are intuitively plausible, Chap. 10 shows that models relying on them fail to provide an adequate account of our visual experience. It is argued that, in light of PP, all four assumptions must be rejected. Instead, we must adopt a view of cognition that closely resembles Kant's account of perception, in which mental categories are imposed on the objects of perception. This is realised in the brain by a two-directional stream of information flow that enables the detection and individuation of objects by projecting categories onto visual data (Chap. 12). This structure can be derived from first principles and is universally shared by all optimal predictive systems (Chap. 11).

Relying on the novel metaphysical framework developed in Part I and the interpretation of PP in Part II, Part III investigates the structural qualities of ordinary objects and the clash between the manifest and scientific images. In Chap. 13, I argue that PP allows us to debunk realism in favour of idealism. This argument builds on two key findings from Part II. First, the computations performed by PP are in relevant respects universal and independent of how ordinary objects are in themselves. Second, the predictive brain emerges as a consequence of any complex system that must survive under evolutionary pressures in a complex and changing environment, rather than as a result of the particular structural qualities of ordinary objects. Since the scientific image does not attribute to ordinary objects the structural qualities we ordinarily associate with them, Chap. 13 concludes that these properties are most likely shaped by our subconscious neural processes.

From this idealist perspective, Chap. 14 argues that we can use the category of an RPP to conceptualise the structural qualities of ordinary objects in a unified manner. Traditional problems concerning the structural qualities of ordinary objects, such as the question of when two objects compose, can then be reformulated using this new category and have straightforward solutions.

The puzzle that is analysed in most detail in this book is the sorites paradox, since this puzzle is often correctly interpreted to be of metaphysical significance, has generated a large body of debate, and its solution can be used to explain a variety of other puzzles concerning ordinary objects. As I explain in Chap. 15, all theories of vagueness to date assume that for any morphing sequence there is some fineness of granularity at which there exists no cut-off, or for which we have no epistemic access to the cut-off. It is shown in detail why we must reject this assumption based on the theory behind PP and on novel experimental findings presented in this book.

The proposed idealist theory generalises to solve the Sellarian concern about how structures at different levels of reality can be irreconcilable (Chap. 16). I show that the clash between different levels of reality can be a consequence of the cogniser's information import into the system, which can vary at different levels of reality. While it is in principle possible to resolve this clash between levels by quantifying the information import of the subject, this calculation is not feasible in most real-world scenarios, since we are unable to fully quantify the information import of complex experimental setups, including the observer's cognitive processes. Consequently, the clash persists, even though different levels of reality belong to one unified world.

From this generalised theory, Chap. 17 evaluates how we must think about some of the puzzles about ordinary objects and discusses how these metaphysical problems come about. With this chapter, the main goal of the book is achieved: providing a general metaphysics of the material world that offers solutions to traditional metaphysical puzzles using a computational neuroscience approach. In the remaining two chapters, I investigate what the implications of my theory are for philosophy in general and consider an application in a different philosophical domain.

As I explain at the outset of this book (Chap. 2), not only problems about the material world but almost all philosophical problems are in some way connected to the clash. This book focuses on the examination of concepts that stand in connection to the material world. However, most philosophical concepts are different. Terms like ‘knowledge’, ‘agency’, ‘normativity’ or ‘free will’ are not clearly connected to the material world and are therefore not analysable by scientific means. Chapter 18 offers a way to understand philosophical problems associated with such concepts from a neuroscience perspective and explains why we must not be eliminativist about these concepts. The explanations I provide are closely connected to the metaphysics I establish in the rest of the book, implying that the proposed theory has a much greater scope than metaphysics alone.

Bibliography

Abu-Mostafa, Y. S., & Psaltis, D. (1987). Optical neural computers. *Scientific American, 256*(3), 88–95.

Åkerman, J. (2009). *Extensions in flux: An essay on vagueness and context sensitivity*. Dep. of Philosophy, Stockholm University.

Allen, M., & Friston, K. J. (2018). From cognitivism to autopoiesis: Towards a computational framework for the embodied mind. *Synthese, 195*(6), 2459–2482.

Andersen, H. K. (2017). Patterns, information, and causation. *The Journal of Philosophy, 114*(11), 592–622.

Anderson, M. L., & Chemero, T. (2013). The problem with brain GUTs: Conflation of different senses of "prediction" threatens metaphysical disaster. *Behavioral and Brain Sciences, 36*(3), 204–205.

Antony, L. (2004). A naturalized approach to the "a priori." *Philosophical Issues, 14*, 1–17.

Aronson, J. L. (1971). On the grammar of "cause." *Synthese, 22*(3–4), 414–430.

Austin, J. L. (1962). *Sense and sensibilia*. Oxford University Press.

Baltieri, M., & Buckley, C. (2019). The dark room problem in predictive processing and active inference, a legacy of cognitivism? In *Artificial Life Conference Proceedings* (pp. 40–47).

Baum, L. F. (1900). *The wonderful wizard of Oz*. George M. Hill Company.

Baz, A. (2020). *Wittgenstein on aspect perception*. Cambridge University Press.

Beale, J. (2020). Wittgenstein's 'grammatical naturalism.' In S. Wuppuluri & N. da Costa (Eds.), *Wittgensteinian (adj.)* (pp. 67–90). Springer International Publishing.

Bealer, G. (1998). Intuition and the autonomy of philosophy. In M. DePaul & W. Ramsey (Eds.), *Rethinking intuition: The psychology of intuition and its role in philosophical inquiry* (pp. 201–239). Rowman & Littlefield.

Beck, O. (2019). Rethinking naïve realism. *Philosophical Studies, 176*(3), 607–633.

Bell, J. (1990). Against "Measurement." *Physics World, 3*(8), 33–41.

Bennett, M., & Hacker, P. (2003). *Philosophical foundations of neuroscience* (Vol. 79). Blackwell Oxford.

Benovsky, J. (2015). From experience to metaphysics: On experience-based intuitions and their role in metaphysics. *Noûs, 49*(4), 684–697.

Benovsky, J. (2016). *Meta-metaphysics*. Springer International Publishing.

Benovsky, J. (2018). *Eliminativism, objects, and persons: The virtues of non-existence*. Taylor & Francis.

Berkeley, G. (1710). Principles of human knowledge. In E. A. Burtt (Ed.), *The English philosophers from Bacon to Mill* (pp. 509–579). Random House.

A. C. Schwaninger, *Cognitive Metaphysics*, Synthese Library 529,
https://doi.org/10.1007/978-3-032-05420-3

Biederman, I. (1987). Recognition-by-components: A theory of human image understanding. *Psychological Review*, *94*(2), 115–147.
Bigaj, T. (2012). *Metaphysics: A guided tour for beginners*.
Bitbol, M. (2010). Reflective metaphysics: Understanding quantum mechanics from a Kantian standpoint. *Philosophica*, *83*(1), 53–83.
Blackburn, S. (1984). *Spreading the word*. Clarendon Press.
Blakemore, C., Fiorentini, A., & Maffei, L. (1972). A second neural mechanism of binocular depth discrimination. *The Journal of Physiology*, *226*(3), 725–749.
Blau, U. (2008). *Die Logik der Unbestimmtheiten und Paradoxien* (Vol. 8). Synchron.
Bones, I., & Raffman, D. (2019). Contextualism and the sorites paradox. In S. Oms & E. Zardini (Eds.), *The sorites paradox*. Cambridge University Press.
Braitenberg, V. (1986). *Vehicles: Experiments in synthetic psychology*. MIT Press.
Brown, J. W. (1998). Foundations of cognitive metaphysics. *Process Studies*, *27*(1), 79–92.
Buffart, H., Leeuwenberg, E. L. J., & Restle, F. (1981). Coding theory of visual pattern completion. *Journal of Experimental Psychology: Human Perception and Performance*, *7*(2), 241–274.
Buhmann, J. M. (2010). Information theoretic model validation for clustering. In *Information Theory Proceedings (ISIT), 2010 IEEE International Symposium on Information Theory* (pp. 1398–1402).
Burge, T. (2014). Perception: Where mind begins. *Philosophy*, *89*(3), 385–403.
Burke, M. B. (1992). Copper statues and pieces of copper: A challenge to the standard account. *Analysis*, *52*(1), 12–17.
Campbell, J., & Cassam, Q. (2014). *Berkeley's puzzle: What does experience teach us?* Oxford University Press.
Cappelen, H. (2012). *Philosophy without intuitions*. Oxford University Press.
Carew, T. J., Castellucci, V. F., & Kandel, E. R. (1971). An analysis of dishabituation and sensitization of the gill-withdrawal reflex in Aplysia. *International Journal of Neuroscience*, *2*(2), 79–98.
Carnap, R. (1950). Empiricism, semantics, and ontology. *Revue Internationale de Philosophie*, *4*(11), 20–40.
Cartwright, N. (1983). *How the laws of physics lie*. Oxford University Press.
Chalmers, D. J. (1996). *The conscious mind: In search of a fundamental theory*. Oxford University Press.
Chalmers, D. J. (2014). Intuitions in philosophy: A minimal defense. *Philosophical Studies*, *171*(3), 535–544.
Chalmers, D. J. (2019). Extended cognition and extended consciousness. In M. Colombo, E. Irvine, & M. Stapleton (Eds.), *Andy clark and his critics* (pp. 9–20). Oxford University Press.
Charlesworth, M. J. (1956). Aristotle's razor. *Philosophical Studies*, *6*(0), 105–112.
Chomsky, N. (1967). *Language and mind*. Cambridge University Press.
Churchland, P. M. (1981). Eliminative materialism and the propositional attitudes. *The Journal of Philosophy*, *78*(February), 67–90.
Churchland, P. S., & Sejnowski, T. J. (1988). Perspectives on cognitive neuroscience. *Science*, *242*(4879), 741–745.
Clark, A. (2013). Whatever next? Predictive brains, situated agents, and the future of cognitive science. *Behavioral and Brain Sciences*, *36*(03), 181–204.
Clark, A. (2015). Radical predictive processing: Radical predictive processing. *The Southern Journal of Philosophy*, *53*, 3–27.
Clark, A. (2016). *Surfing uncertainty: Prediction, action, and the embodied mind*. Oxford University Press.
Clark, A. (2017). How to knit your own Markov blanket: Resisting the second law with metamorphic minds. *Philosophy and Predictive Processing*, *3*.
Clark, A., & Chalmers, D. (1998). The extended mind. *Analysis*, *58*(1), 7–19.
Clarke, T. (1965). Seeing surfaces and physical objects. In M. Black (Ed.), *Philosophy in America* (pp. 98–114). Cornell University Press.
Climenhaga, N. (2018). Intuitions are used as evidence in philosophy. *Mind*, *127*(505), 69–104.

Cohen, J. (2015). Perceptual representation, veridicality, and the interface theory of perception. *Psychonomic Bulletin & Review*, *22*(6), 1512–1518.

Colombo, M., & Wright, C. (2018). First principles in the life sciences: the free-energy principle, organicism, and mechanism. *Synthese*, *198*(14), 3463–3488.

Cooper, J. (2007). *Cognitive dissonance: 50 years of a classic theory*. Sage.

Cover, T. M., & Thomas, J. A. (2006). *Elements of information theory* (2nd ed.). Wiley.

Crane, T. (2015). *The mechanical mind: A philosophical introduction to minds, machines and mental representation*. Routledge.

Creutzfeldt, O. D. (1977). Generality of the functional structure of the neocortex. *Naturwissenschaften*, *64*(10), 507–517.

d'Holbach, B. (1770). *The System of nature or laws of the moral and physical world*. (H. D. Robinson, Trans.) (Vols. 1–2, Vol. 1). J. P. Mendum.

de La Mettrie, J. O. (1748). *Man a machine*. (G. Bussey, Trans.). Open Court Publishing Company.

De Pittà, M., & Berry, H. (2019). A Neuron–Glial perspective for computational neuroscience. In M. De Pittà & H. Berry (Eds.), *Computational glioscience* (pp. 3–35). Springer International Publishing.

Decock, L. (2018). Cognitive metaphysics. *Frontiers in Psychology*, *9*, 1700.

Dennett, D. (1975). Why the law of effect will not go away. *Journal for the Theory of Social Behaviour*, *5*, 169–187.

Dennett, D. (1991). Real patterns. *The Journal of Philosophy*, *88*(1), 27–51.

Dennett, D. (1998). Can machines think? In D. J. Levitin (Ed.), *Foundations of cognitive psychology: Core readings*. MIT Press.

Dennett, D. (2013). Kinds of things – Towards a bestiary of the manifest image. In D. Ross, J. Ladyman, & H. Kincaid (Eds.), *Scientific metaphysics*. Oxford University Press.

Descartes, R. (1641). *Meditations on first philosophy*. Broadview Press.

Descartes, R. (1988). *The philosophical writings of descartes*. (J. Cottingham, R. Stoothoff, & D. Murdoch, Trans.) (Vol. I). Cambridge University Press.

Deutsch, M. (2009). Experimental philosophy and the theory of reference. *Mind & Language*, *24*(4), 445–466.

Deutsch, M. (2010). Intuitions, counter-examples, and experimental philosophy. *Review of Philosophy and Psychology*, *1*(3), 447–460.

Dewey, J. (1900). Some stages of logical thought. *The Philosophical Review*, *9*(5), 465–489.

Douven, I., & Decock, L. (2010). Identity and similarity. *Philosophical Studies*, *151*(1), 59–78.

Dowe, P. (2000). *Physical causation*. Cambridge University Press.

Dworkin, R. (2011). *Justice for hedgehogs*. Belknap Press of Harvard University Press.

Eddington, A. S. (1928). *The nature of the physical world*. New York: Macmillan.

Egerton, K. (2016). Getting off the Inwagen: a critique of quinean metaontology. *Journal for the History of Analytical Philosophy*, *4*(6).

Egerton, K. (2020). Quine's metametaphysics. In R. Bliss & J. T. M. Miller (Eds.), *The Routledge handbook of metametaphysics* (pp. 49–60). Routledge.

Engel, A. K., Fries, P., & Singer, W. (2001). Dynamic predictions: Oscillations and synchrony in top–down processing. *Nature Reviews Neuroscience*, *2*(10), 704–716.

Esfeld, M. (1999a). Quantum holism and the philosophy of mind. *Journal of Consciousness Studies*, *6*(1), 23–38.

Esfeld, M. (1999b). Holism in cartesianism and in today's philosophy of physics. *Journal for General Philosophy of Science*, *30*(1), 17-36.

Ewing, A. C. (1938). *A short commentary on Kant's critique of pure reason*. The University of Chicago Press.

Falkenstein, L. (1995). *Kant's intuitionism*. Toronto University Press.

Feldman, J. (2013). Tuning your priors to the world. *Topics in Cognitive Science*, *5*(1), 13–34.

Feynman, R. P. (1963). *Six easy pieces: Essentials of physics explained by its most brilliant teacher*. Basic Books.

FitzPatrick, W. J. (2015). Debunking evolutionary debunking of ethical realism. *Philosophical Studies*, *172*(4), 883–904.

Fodor, J. A. (1998). *Concepts: Where cognitive science went wrong*. Clarendon Press.

Franklin, J. (2001). *The science of conjecture: Evidence and probability before Pascal*. Johns Hopkins University Press.

French, S. (2019). There are no such things as ordinary objects. In J. Cumpa & B. Brewer (Eds.), *The nature of ordinary objects* (pp. 260–273). Cambridge University Press.

French, S., & Redhead, M. (1988). Quantum physics and the identity of indiscernibles. *The British Journal for the Philosophy of Science*, *39*(2), 233–246.

Friston, K. (2003). Learning and inference in the brain. *Neural Networks*, *16*(9), 1325–1352.

Friston, K. (2005). A theory of cortical responses. *Philosophical Transactions of the Royal Society: Biological Sciences*, *360*(1456), 815–836.

Friston, K. (2008). Hierarchical models in the brain. *PLoS Computational Biology*, *4*(11), e1000211.

Friston, K. (2009). The free-energy principle: A rough guide to the brain? *Trends in Cognitive Sciences*, *13*(7), 293–301.

Friston, K. (2010). The free-energy principle: A unified brain theory? *Nature Reviews Neuroscience*, *11*(2), 127–138.

Friston, K. (2012a). The history of the future of the Bayesian brain. *NeuroImage*, *62*(2), 1230–1233.

Friston, K. (2012b). A free energy principle for biological systems. *Entropy*, *14*(11), 2100–2121.

Friston, K. (2013a). Active inference and free energy. *Behavioral and Brain Sciences*, *36*(3), 212–213.

Friston, K. (2013b). Life as we know it. *Journal of the Royal Society Interface*, *10*(86), 20130475.

Friston, K., FitzGerald, T., Rigoli, F., Schwartenbeck, P., O'Doherty, J., & Pezzulo, G. (2016). Active inference and learning. *Neuroscience & Biobehavioral Reviews*, *68*, 862–879.

Friston, K., FitzGerald, T., Rigoli, F., Schwartenbeck, P., & Pezzulo, G. (2017). Active inference: A process theory. *Neural Computation*, *29*(1), 1–49.

Friston, K., & Frith, C. D. (2015). Active inference, communication and hermeneutics. *Cortex*, *68*, 129–143.

Friston, K., & Kiebel, S. (2009). Predictive coding under the free-energy principle. *Philosophical Transactions of the Royal Society B: Biological Sciences*, *364*(1521), 1211–1221.

Friston, K., Kilner, J., & Harrison, L. (2006). A free energy principle for the brain. *Journal of Physiology-Paris*, *100*(1–3), 70–87.

Friston, K., & Stephan, K. E. (2007). Free-energy and the brain. *Synthese*, *159*(3), 417–458.

Friston, K., Thornton, C., & Clark, A. (2012). Free-energy minimization and the dark-room problem. *Frontiers in Psychology*, *3*, 130.

Frith, C. (2007). *Making up the mind: How the brain creates our mental world*. Blackwell Publishing.

Garcia, J., & Koelling, R. A. (1996). *Relation of cue to consequence in avoidance learning*. University of Chicago Press.

Gärdenfors, P. (1996). Mental representation, conceptual spaces and metaphors. *Synthese*, *106*, 21–47.

Gärdenfors, P. (2000). *Conceptual spaces: The geometry of thought*. MIT Press.

Garner, W. R. (1974). *The processing of information and structure*. Erlbaum.

Geach, P. T. (1967). Identity. *Review of Metaphysics*, *21*(1), 2–12.

Geman, S., Bienenstock, E., & Doursat, R. (1992). Neural networks and the bias/variance dilemma. *Neural Computation*, *4*(1), 1–58.

Gershman, S. J. (2019). The generative adversarial brain. *Frontiers in Artificial Intelligence*, *2*, 18.

Gettier, E. (1963). Is justified true belief knowledge? *Analysis*, *23*, 121–123.

Gibson, J. (1966). *The senses considered as perceptual systems*. Waveland Press, Inc.

Gibson, J. (1969). *The senses considered as perceptual systems*. Houghton Mifflin.

Gibson, J. (1979). *The ecological approach to visual perception*. Psychology Press.

Glock, H.-J. (1997). Kant and Wittgenstein: Philosophy, necessity and representation. *International Journal of Philosophical Studies*, *5*(2), 22.

Glock, H.-J. (2003). Quine and Davidson on language. Cambridge University Press.

Glock, H.-J. (2010). Concepts; between the subjective and the objective. In J. Hacker & P. Hacker (Eds.), *Mind, method and morality: Essays in honour of Anthony Kenny* (pp. 306–329). Oxford University Press.

Glock, H.-J. (2011). Evolutionäre Erkenntnistheorie. In H.-U. Reyer & P. Schmid-Hempel (Eds.), *Darwins langer Arm – Evolutionstheorie heute*. vdf Hochschulverlag AG.

Glock, H.-J. (2022). Strawson and non-revisionary naturalism. In *The Routledge handbook of liberal naturalism*. Routledge.

Goldman, A. I. (1989). Metaphysics, mind, and mental science. *Philosophical Topics*, *17*(1), 131–145.

Goldman, A. I. (2007). Philosophical intuitions: Their target, their source, and their epistemic status. *Grazer Philosophische Studien*, *4*, 1–26.

Goldman, A. I., & Prust, J. (1998). Philosophical theory and intuitional evidence. In J. Pust & W. Ramsey (Eds.), *Rethinking intuition: The psychology of intuition and its role in philosophical inquiry* (pp. 179–197). Rowman & Littlefield Publishers.

Graff Fara, D. (2000). Shifting sands: An interest-relative theory of vagueness. *Philosophical Topics*, *28*(1), 45–81.

Green, E. J. (2019). A theory of perceptual objects. *Philosophy and Phenomenological Research*, *99*(3), 663–693.

Green, E. J. (2021). Binding and differentiation in multisensory object perception. *Synthese*, *198*(5), 4457–4491.

Greenough, P. (2017). Neutralism and the observational sorites paradox. Synthese.

Gregory, R. (1970). *The intelligent eye*. Weidenfeld and Nicolson.

Gregory, R. (1972). Seeing as thinking: An active theory of perception. *Times Literary Supplement*, 11.

Gribbin, J. (1984). *In search of Schrödinger's cat – Quantum physics and reality*. Bantam Books.

Haack, S. (1993). The two faces of Quine's naturalism. *Synthese*, *94*(3), 335–356.

Harman, G. (2010). I am also of the opinion that materialism must be destroyed. *Environment and Planning D: Society and Space*, *28*(5), 772–790.

Hassenstein, B., & Reichardt, W. (1956). Systemtheoretische Analyse der Zeit-, Reihenfolgen- und Vorzeichenauswertung bei der Bewegungsperzeption des Rüsselkäfers Chlorophanus. *Zeitschrift für Naturforschung B*, *11*(9–10), 513–524.

Hebb, D. O. (1949). *The organization of behavior*. Wiley.

Heck, R. (2003). Semantic accounts of vagueness. In J. C. Beall (Ed.), *Liars and heaps*. Oxford University Press.

Heisenberg, W. (1958). *Physics and philosophy: The revolution in modern science*. Harper & Brothers Publishers.

Heller, M. (1990). *The ontology of physical objects: Four-dimensional hunks of matter*. Cambridge University Press.

Hertz, J. (1991). *Introduction to the theory of neural computation*. Addison-Wesley Publishing Company.

Higgins, A. A. (2014). *The nature of intuitions and their role in material object metaphysics*. University of Illinois, Urbana-Champaign.

Hilpinen, R. (1996). On some formulations of realism, or how many objects are there in the world? In R. S. Cohen, R. Hilpinen, & Q. Renzong (Eds.), *Realism and Anti-Realism in the Philosophy of Science: Beijing International Conference, 1992* (pp. 1–10). Springer Netherlands.

Hinton, G. (2007). To recognize shapes, first learn to generate images. In *Progress in brain research* (Vol. 165, pp. 535–547). Elsevier.

Hinton, G., Dayan, P., Frey, B., & Neal, R. (1995). The "wake-sleep" algorithm for unsupervised neural networks. *Science*, *268*(5214), 1158–1161.

Hirsch, E. (2002). Quantifier variance and realism. *Philosophical Issues*, *12*, 51–73.

Hobbes, T. (1655). *De Corpore*. (W. Molesworth, Trans.) (Vol. II). Scientia-Verlag.

Hoffman, D. D. (2019). *The case against reality: How evolution hid the truth from our eyes*. W.W. Norton & Co.

Hoffman, D. D., Singh, M., & Prakash, C. (2015). The interface theory of perception. *Psychonomic Bulletin & Review*, *22*(6), 1480–1506.
Hoffman, J., & Rosenkrantz, G. S. (1997). *Substance: Its nature and existence*. Routledge.
Hofweber, T. (2019a). Empirical evidence and the metaphysics of ordinary objects. In J. Cumpa & B. Brewer (Eds.), *The nature of ordinary objects* (pp. 27–47). Cambridge University Press.
Hofweber, T. (2019b). Idealism and the harmony of thought and reality. *Mind*, *128*(511), 699–734.
Hohwy, J. (2013). *The predictive mind*. Oxford University Press.
Hohwy, J. (2016). The self-evidencing brain. *Noûs*, *50*(2), 259–285.
Hohwy, J., Roepstorff, A., & Friston, K. (2008). Predictive coding explains binocular rivalry: An epistemological review. *Cognition*, *108*(3), 687–701.
Hohwy, J., & Seth, A. (2020). Predictive processing as a systematic basis for identifying the neural correlates of consciousness. *Philosophy and the Mind Sciences*, *1*(II), 3.
Hopfield, J. J. (1982). Neural networks and physical systems with emergent collective computational abilities. *Proceedings of the National Academy of Sciences*, *79*(8), 2554–2558.
Horgan, T. (1993). On what there isn't. *Philosophy and Phenomenological Research*, *53*(3), 693–700.
Horgan, T. (1994). Robust vagueness and the forced-march sorites paradox. *Philosophical Perspectives*, *8*, 159–188.
Horgan, T., & Potrč, M. (2008). *Austere realism: Contextual semantics meets minimal ontology*. MIT Press.
Horn, K. S. V. (2003). Constructing a logic of plausible inference: A guide to Cox's theorem. *International Journal of Approximate Reasoning*, 22.
Hornik, K., Stinchcombe, M., & White, H. (1989). Multilayer feedforward networks are universal approximators. *Neural Networks*, *2*(5), 359–366.
Howard, D. (1989). Holism, separability, and the metaphysical implications of the Bell experiments. In J. Cushing & E. McMullin (Eds.), *Philosophical consequences of quantum theory: Reflections on Bell's theorem*. University of Notre Dame Press.
Hubel, D. H., & Wiesel, T. N. (1962). Receptive fields, binocular interaction and functional architecture in the cat's visual cortex. *The Journal of Physiology*, *160*(1), 106–154.
Hubel, D. H., & Wiesel, T. N. (1968). Receptive fields and functional architecture of monkey striate cortex. *Journal of Physiology*, *195*, 215–243.
Hubel, D. H., & Wiesel, T. N. (1974). Uniformity of monkey striate cortex: A parallel relationship between field size, scatter, and magnification factor. *Journal of Comparative Neurology*, *158*(3), 295–305.
Huberman, B., & Hogg, T. (1984). Adaptation and self-repair in parallel computing structures. *Physical Review Letters*, *52*(12), 1048–1051.
Huberman, B., & Hogg, T. (1986). Complexity and Adaptation. *Physica D*, *2*(1–3), 376–384.
Hume, D. (1739). *A treatise of human nature*. Oxford University Press.
Humphreys, P. (2013). Scientific ontology and speculative ontology. In D. Ross, J. Ladyman, & H. Kincaid (Eds.), *Scientific metaphysics* (pp. 51–78). Oxford University Press.
Husmann, J., & Näger, P. M. (2018). Physical composition by bonding. In L. Jansen & P. M. Näger (Eds.), *Peter van Inwagen* (Vol. 4, pp. 65–96). Springer International Publishing.
Hyde, D., & Raffman, D. (1997). Sorites paradox. In *The Stanford encyclopedia of philosophy*. Metaphysics Research Lab, Stanford University. Retrieved January 21, 2020, from https://plato.sydney.edu.au/archives/sum2018/entries/sorites-paradox/
Isaac, A. M. (2018). Computational thought from Descartes to Lovelace. *The Routledge handbook of the computational mind* (pp. 9–22). Routledge.
Jackson, F. (1982). Epiphenomenal Qualia. *The Philosophical Quarterly*, *32*(127), 127–136.
Jäkel, F., Singh, M., Wichmann, F., & Herzog, M. H. (2016). An overview of quantitative approaches in Gestalt perception. *Vision Research*, *126*, 3–8.
James, W., & Stein, C. (1961). Estimation with quadratic loss. In J. Neyman (Ed.), *Proceedings of the Fourth Berkeley Symposium* (pp. 361–380). University of California Press.
Jastrow, J. (1899). The mind's eye. *Popular Science Monthly*, *54*, 299–312.

Jaynes, E. T. (1986). Bayesian methods: General background. In J. H. Justice (Ed.), *Maximum entropy and Bayesian methods in applied statistics* (pp. 1–25). Cambridge University Press.

Jaynes, E. T. (1989). Clearing up mysteries - The original goal. In J. Skilling (Ed.), *Maximum entropy and Bayesian methods: Cambridge, England, 1988* (pp. 1–27). Springer Science+Business Media.

Jaynes, E. T. (2003). *Probability theory: The logic of science*. Cambridge University Press.

Johansson, L.-G. (2011). Kant versus Quine: Transcendentalism or naturalism? *Uppsala Philosophical Studies, 54*, 115–128.

Kaaronen, R. O. (2018). A theory of predictive dissonance: Predictive processing presents a new take on cognitive dissonance. *Frontiers in Psychology, 9*, 2218.

Kahneman, D. (2011). *Thinking, fast and slow*. Macmillan.

Kant, I. (1781/1787). *Critique of pure reason*. (W. S. Pluhar, Trans., 1996). Hackett Publishing Company.

Kant, I. (1781). Letter to Marcus Herz. Trans. A. Zweig, 1999. In *Correspondence* (pp. 180–182). Cambridge University Press.

Karni, A., Tanne, D., Rubenstein, B. S., Askenasy, J. J., & Sagi, D. (1994). Dependence on REM sleep of overnight improvement of a perceptual skill. *Science, 265*(5172), 679–682.

Keefe, R. (2000). *Theories of vagueness*. Cambridge University Press.

Keefe, R. (2003). Context, vagueness, and the sorites. In J. C. Beall (Ed.), *Liars and heaps* (pp. 73–83). Oxford University Press.

Keefe, R. (2007). Vagueness without context change. *Mind, 116*(462), 275–292.

Keefe, R. (2011). Phenomenal sorites paradoxes and looking the same. *Dialectica, 65*(3), 327–344.

Kemp, C., Perfors, A., & Tenenbaum, J. B. (2007). Learning overhypotheses with hierarchical Bayesian models. *Developmental Science, 10*(3), 307–321.

Kenny, A. (1984). *The legacy of Wittgenstein*. Basil Blackwell.

Kersten, D., Mamassian, P., & Yuille, A. (2004). Object perception as Bayesian inference. *Annual Review of Psychology, 55*(1), 271–304.

Keskinen, A. (2012). Quine on objects: Realism or anti-realism? *Theoria, 78*(2), 128–145.

Kiefer, A., & Hohwy, J. (2020). Representation in the prediction error minimization framework. In S. Robins, J. Symons, & P. Calvo (Eds.), *The Routledge companion to philosophy of psychology* (2nd ed., pp. 384–409). Routledge, Taylor & Francis Group.

King, D. B., & Wertheimer, M. (2005). *Max Wertheimer and Gestalt theory*. Transaction Publishers.

Kirchhoff, M. (2019). *Extended consciousness and predictive processing: A third wave view*. Routledge.

Kirchhoff, M., & Kiverstein, J. (2019). How to determine the boundaries of the mind: A Markov blanket proposal. *Synthese, 198*(8), 4791–4810.

Kirchhoff, M., Parr, T., Palacios, E., Friston, K., & Kiverstein, J. (2018). The Markov blankets of life: Autonomy, active inference and the free energy principle. *Journal of the Royal Society Interface, 15*(138), 20170792.

Kirillov, A., Mintun, E., Ravi, N., Mao, H., Rolland, C., Gustafson, L., et al. (2023). Segment anything. *2023 IEEE/CVF International Conference on Computer Vision (ICCV)* (pp. 3992–4003).

Koffka, K. (1935). *Principles of gestalt psychology*. Brace & World.

Köhler, W. (1947). *Gestalt psychology*. Liveright.

Kohonen, T. (1978). *Associative memory: A system theoretic approach*. Springer-Verlag.

Koksvik, O. (2020). *Intuition as conscious experience*. Routledge.

Kolmogorov, A. N. (1963). On tables of random numbers. *Sankhyā: The Indian Journal of Statistics, Series A, 25*, 369–376.

Korman, D. Z. (2011). Ordinary objects. In *The Stanford encyclopedia of philosophy* (Summer 2021). Metaphysics Research Lab, Stanford University. Retrieved September 27, 2022, from https://plato.sydney.edu.au/archives/sum2021/entries/ordinary-objects/

Korman, D. Z. (2014). Debunking perceptual beliefs about ordinary objects. *Philosophers, 14*(13), 21.

Korman, D. Z. (2015). *Objects: Nothing out of the ordinary*. Oxford University Press.
Kornblith, H. (1998). The role of intuitions in philosophical enquiry: An account with no unnatural ingredients. In M. DePaul & W. Ramsey (Eds.), *Rethinking intuition: The psychology of intuition and its role in philosophical inquiry* (pp. 129–41). Rowman & Littlefield Publishers.
Kornblith, H. (2002). *Knowledge and its place in nature*. Oxford University Press.
Körner, S. (1960). *The philosophy of mathematics: An introductory essay*. Hutchinson University Library.
Koster-Hale, J., & Saxe, R. (2013). Theory of mind: A neural prediction problem. *Neuron, 79*(5), 836–848.
Kraut, R. (2020). Rudolf Carnap: Pragmatist and expressivist about ontology. In R. Bliss & J. T. M. Miller (Eds.), *The Routledge handbook of metametaphysics* (pp. 32–48). Routledge.
Kriegel, U. (2008). Composition as a secondary quality. *Pacific Philosophical Quarterly*, *89*(3), 359–383.
Ladyman, J. (2009). What does it mean to say that a physical system implements a computation? *Theoretical Computer Science*, *410*(4–5), 376–383.
Ladyman, J., & Ross, D. (2007). *Every thing must go: Metaphysics naturalized*. Oxford University Press.
Ladyman, J., & Ross, D. (2013). The world in the data. In D. Ross, J. Ladyman, & H. Kincaid (Eds.), *Scientific metaphysics* (pp. 108–150). Oxford University Press.
Laplace, P.-S. (1812). *Pierre-Simon Laplace philosophical essay on probabilities: Translated from the fifth French edition of 1825*. Springer Science & Business Media.
Laughlin, R. B. (2005). *A different universe: Reinventing physics from the bottom down*. Basic Books.
Leahey, T. H. (2017). *A history of psychology: From antiquity to modernity*. Routledge.
Leeuwenberg, E. L. J. (1968). *Structural information of visual patterns: An efficient coding system in perception*. The Hague: Mouton.
Leeuwenberg, E. L. J., & Van der Helm, P. A. (2013). *Structural information theory: The simplicity of visual form*. Cambridge University Press.
Lewis, D. (1976). Survival and identity. In A. Rorty (Ed.), *The identities of persons* (pp. 17–40). University of California Press.
Lewis, D. (1986a). *On the plurality of worlds* (Vol. 322). Oxford Blackwell.
Lewis, D. (1986b). *Philosophical papers*. Oxford University Press.
Lewis, D. (1991). *Parts of classes*. Basil Blackwell.
Libet, B., Gleason, C. A., Wright, E. W., & Pearl, D. K. (1983). Time of conscious intention to act in relation to onset of cerebral activity (readiness-potential) the unconscious initiation of a freely voluntary act. *Brain, 106*(3), 623–642.
Locke, J. (1847). *An essay concerning human understanding*. Kay & Troutman.
Loux, M. J. (2006). *Metaphysics: A contemporary introduction* (3rd ed.). Routledge.
Lowe, E. J. (1989). What is a criterion of identity?. *The Philosophical Quarterly*, *29*, 1–21.
Lowe, E. J. (1997). Objects and criteria of identity. In B. Hale & C. Wright (Eds.), *A companion to the philosophy of language* (pp. 613–633). Blackwell.
Lowe, E. J. (2005). How are ordinary objects possible?: *Monist*, *88*(4), 510–533.
Lycan, W. G. (1991). Homuncular functionalism meet PDP. In W. M. Ramsey, S. P. Stich, & D. E. Rumelhart (Eds.), *Philosophy and connectionist theory*. Lawrence Erlbaum Associates, Publishers.
Macarthur, D. (2017). Liberal naturalism and philosophy of the manifest image. In A. Gare & W. Hudson (Eds.), *For a new naturalism* (pp. 50–65). Telos Press.
Machina, K. F. (1972). Kant, Quine, and human experience. *The Philosophical Review*, *81*(4), 484–497.
Machina, K. F. (1976). Truth, belief, and vagueness. *Journal of Philosophical logic*, 47–78.
Mann, S. F., Pain, R., & Kirchhoff, M. D. (2022). Free energy: A user's guide. *Biology & Philosophy*, *37*(4), 33.
Mark, J. T., Marion, B. B., & Hoffman, D. D. (2010). Natural selection and veridical perceptions. *Journal of Theoretical Biology*, *266*(4), 504–515.

Markosian, N. (1998). Brutal composition. *Philosophical Studies, 92*(3), 211–249.
Marr, D. (2010). *Vision*. The MIT Press.
Martin, M. G. F. (2006). On being alienated. In T. S. Gendler & J. Hawthorne (Eds.), *Perceptual experience* (pp. 354–410). Clarendon Press:
Martínez, M. (2019). Usefulness drives representations to truth: A family of counterexamples to Hoffman's interface theory of perception. *Grazer Philosophische Studien, 96*(3), 319–341.
Maslin, K. (2001). *An introduction to the philosophy of mind*. Polity.
Maturana, H., & Varela, F. (1984). *The tree of knowledge: The biological roots of human understanding*. Shambhala Publications, Inc.
Maudlin, T. (1998). Part and whole in quantum mechanics. In E. Castellani (Ed.), *Interpreting bodies* (pp. 46–60). Princeton University Press.
Maudlin, T. (2007). *The metaphysics within physics*. Oxford University Press.
Maudlin, T. (2019). *Philosophy of physics: Quantum theory*. Princeton University Press.
Maund, B. (2011). Colour eliminativism. In L. Nolan (Ed.), *Primary and secondary qualities: The historical and ongoing debate*. Oxford University Press.
Mausfeld, R. (2015). Notions such as "truth" or "correspondence to the objective world" play no role in explanatory accounts of perception. *Psychonomic Bulletin & Review, 22*(6), 1535–1540.
McAllister, J. W. (2011). What do patterns in empirical data tell us about the structure of the world? *Synthese, 182*(1), 73–87.
McDaniel, K. (2020). *This is metaphysics: An introduction*. Wiley, Blackwell.
McGrath, M. (2005). No objects, no problem? *Australasian Journal of Philosophy, 83*(4), 457–486.
Meinong, A. (1904). Über Gegenstandstheorie. In *Untersuchungen zur Gegenstandstheorie und Psychologie*. Barth.
Mermin, D. (1989). What's wrong with this pillow? *Physics Today, 42*(4), 9–11.
Merricks, T. (2001). *Objects and persons*. Oxford University Press.
Merzenich, M. M., Kaas, J., Wall, J., Nelson, R., Sur, M., & Felleman, D. (1983). Topographic reorganization of somatosensory cortical areas 3b and 1 in adult monkeys following restricted deafferentation. *Neuroscience, 8*(1), 33–55.
Metzinger, T. (2009). *The ego tunnel: The science of the mind and the myth of the self*. Basic Books.
Moore, A. W. (2012). *The evolution of modern metaphysics: Making sense of things*. Cambridge University Press.
Mumford, D. (2000). The dawning of the age of stochasticity. *Mathematics: Frontiers and Perspectives, 9*(11), 197–218.
Nado, J. (2014). Why intuition? *Philosophy and Phenomenological Research, 89*(1), 15–41.
Nagel, E. (1961). *The structure of science*. Harcourt, Brace, and World.
Ney, A. (2012). Neo-positivist metaphysics. *Philosophical Studies, 160*(1), 53–78.
O'Callaghan, C. (2016). Objects for multisensory perception. *Philosophical Studies, 173*(5), 1269–1289.
O'Leary-Hawthorne, J., & Michael, M. (1996). Compatibilist semantics in metaphysics: A case study. *Australasian Journal of Philosophy, 74*(1), 117–134.
Oppenheim, P., & Putnam, H. (1958). Unity of science as a working hypothesis. *Minnesota Studies in the Philosophy of Science, 2*, 3–36.
Palmer, S., & Rock, I. (1994). Rethinking perceptual organization: The role of uniform connectedness. *Psychonomic Bulletin & Review, 1*(1), 29–55.
Papazoglou, A. (2019). The idealist challenge to naturalism. In *Responses to naturalism* (pp. 98–119). Routledge.
Paton, H. J. (1936). *Kant's metaphysic of experience*. George Allen and Unwin.
Pearl, J. (1988). *Probabilistic reasoning in intelligent systems: Networks of plausible inference*. Morgan Kaufmann Publishers.
Penrose, R. (1989). *The emperor's new mind*. Oxford University Press.
Pereboom, D. (1991). Is Kant's transcendental philosophy inconsistent? *History of Philosophy Quarterly, 8*(4), 357–372.
Petersen, S. (2019). Composition as pattern. *Philosophical Studies, 176*(5), 1119–1139.

Petrovici, M. A. (2016). *Form versus function: Theory and models for neuronal substrates*. Springer International Publishing.
Phemister, P. (2006). *The rationalists: Descartes, Spinoza and Leibniz*. Polity.
Pinker, S. (1997). *How the mind works*. Penguin Books.
Pippin, R. B. (2005). *The persistence of subjectivity: On the Kantian aftermath*. Cambridge University Press.
Plato, & Waterfield, R. (2002). *Phaedrus*. Oxford University Press.
Pouget, A., Beck, J. M., Ma, W. J., & Latham, P. E. (2013). Probabilistic brains: Knowns and unknowns. *Nature Neuroscience*, *16*(9), 1170–1178.
Powers, W. T. (1973). *Behavior: The control of perception* (Vol. 85). Aldine.
Powers, W. T. (1978). Quantitative analysis of purposive systems: Some spadework at the foundations of scientific psychology. *Psychological Review*, *85*(5), 417.
Price, H. (2009). Metaphysics after Carnap: The ghost who walks? In D. J. Chalmers, D. Manley, & R. Wasserman (Eds.), *Metametaphysics: New essays on the foundations of ontology* (pp. 320–346). Clarendon Press/Oxford University Press.
Priest, G. (2003). A site for sorites. In J. C. Beall (Ed.), *Liars and heaps* (pp. 9–23). Oxford University Press.
Ptak, R., Turri, F., & Doganci, N. (2020). Object recognition and visual agnosia. In *Reference module in neuroscience and biobehavioral psychology*. Elsevier.
Pust, J. (2012). Intuition. In *The Stanford encyclopedia of philosophy* (Summer 2017). Metaphysics Research Lab, Stanford University. Retrieved October 2, 2022, from https://seop.illc.uva.nl/entries/intuition/
Putnam, H. (1981). *Reason, truth, and history*. Cambridge University Press.
Putnam, H. (1985). *The many faces of realism*. Open Court.
Putnam, H. (1987). Truth and convention: On Davidson's refutation of conceptual relativism. *Dialectica*, *41*(1–2), 69–77.
Putnam, H. (1988). *Representation and reality*. MIT Press.
Putnam, H. (1990). *Realism with a human face*. Harvard University Press.
Quine, W. V. O. (1948). On what there is. *The Review of Metaphysics*, *2*(5), 21–38.
Quine, W. V. O. (1960). *Word and object*. Technology Press and Wiley.
Quine, W. V. O. (1963). *From a logical point of view*. Harper and Row.
Quine, W. V. O. (1969). *Ontological relativity: And other essays*. Columbia University Press.
Quine, W. V. O. (1970). *Philosophy of logic*. Harvard University Press.
Quine, W. V. O. (1981). *Theories and things*. Harvard University Press.
Quine, W. V. O. (1993). In praise of observation sentences. *The Journal of Philosophy*, *90*(3), 107–116.
Quine, W. V. O. (1995). Naturalism; Or, living within one's means. *Dialectica*, *49*(2–4), 251–263.
Raffman, D. (1994). Vagueness without paradox. *The Philosophical Review*, *103*(1), 35.
Raffman, D. (1996). Vagueness and context-relativity. *Philosophical Studies*, *81*(2/3), 175–192.
Raffman, D. (2014). *Unruly words: A study of vague language*. Oxford University Press.
Ramón y Cajal, S. (1911). Histologie du système nerveux de l'homme et des vertébrés. *Maloine, Paris*, *2*, 153–173.
Ramstead, M. J. D., Badcock, P. B., & Friston, K. J. (2018). Answering Schrödinger's question: A free-energy formulation. *Physics of Life Reviews*, *24*, 1–16.
Rao, R. P., & Ballard, D. H. (1999). Predictive coding in the visual cortex: A functional interpretation of some extra-classical receptive-field effects. *Nature Neuroscience*, *2*(1), 79–87.
Reichardt, W. (1990). Movement detection and figure-ground discrimination. In L. Deecke, J. C. Eccles, & V. B. Mountcastle (Eds.), *From neuron to action* (pp. 267–276). Springer.
Reicher, M. (2019). Nonexistent objects. In E. N. Zalta (Ed.), *The Stanford encyclopedia of philosophy* (Winter 2019). Metaphysics Research Lab, Stanford University. Retrieved September 27, 2022, from https://plato.stanford.edu/archives/win2019/entries/nonexistent-objects/
Rescher, N. (2016). On kinds of things and cognitive idealism. *Idealistic Studies*, *46*(1), 1–17.
Restle, F. (1979). Coding theory of the perception of motion configurations. *Psychological Review*, *86*(1), 1–24.

Ribeiro, C. (2015). Unjustified criticism of metaphysics. *Lato Sensu, Revue de la Société de philosophie des sciences*, *2*(1), 1–13.

Rickless, S. C. (1997). Locke on primary and secondary qualities. *Pacific Philosophical Quarterly*, *78*(3), 297–319.

Rockel, A. J., Hiorns, R. W., & Powell, T. P. (1980). The basic uniformity in structure of the neocortex. *Brain: A Journal of Neurology*, *103*(2), 221–244.

Rosen, G., & Dorr, C. (2002). Composition as a fiction. In R. M. Gale (Ed.), *The Blackwell guide to metaphysics* (pp. 151–174). Blackwell Publishing Ltd.

Rosenberg, A. (2011). *The atheist's guide to reality: Enjoying life without illusions*. WW Norton & Company.

Rosenberg, A. (2017). Strong scientism and its research agenda. In M. Boudry & M. Pigliucci (Eds.), *Science unlimited? The challenges of scientism*. The University of Chicago Press.

Russell, B. (1905). On denoting. *Mind*, *14*, 479–493.

Russell, B. (1912). *The problems of philosophy*. Oxford University Press.

Russell, B. (1948). *Human knowledge: Its scope and limits*. George Allen & Unwin.

Sainsbury, R. M. (1990). *Concepts without boundaries*. Inaugural lecture published by the King's College London Department of Philosophy.

Sanford, D. H. (1976). Competing semantics of vagueness: Many values versus supertruth. *Synthese*, *33*, 195–210.

Saporiti, K. (2004). Berkeley's perceptual realism. In R. Schumacher (Ed.), *Perception and reality. From Descartes to the present* (pp. 188–213). Mentis.

Schaffer, J. (2003). Is there a fundamental level? *Nous*, *37*(3), 498–517.

Schaffer, J. (2009). On what grounds what. In D. J. Chalmers, D. Manley, & R. Wasserman (Eds.), *Metametaphysics: New essays on the foundations of ontology* (pp. 347–383). Clarendon Press/Oxford University Press.

Schlicht, T., & Dolega, K. (2021). You can't always get what you want: Predictive processing and consciousness. *Philosophy and the Mind Sciences*, *2*.

Schlick, M. (1948). Positivism and realism. *Synthese*, *7*(6-B), 478–505.

Scholl, B. J. (2007). Object persistence in philosophy and psychology. *Mind & Language*, *22*(5), 563–591.

Schrimpf, M., Blank, I. A., Tuckute, G., Kauf, C., Hosseini, E. A., Kanwisher, N., et al. (2021). The neural architecture of language: Integrative modeling converges on predictive processing. *Proceedings of the National Academy of Sciences*, *118*(45), e2105646118.

Schrödinger, E. (1935). Die gegenwärtige Situation in der Quantenmechanik. *Naturwissenschaften*, *23*(48), 807–812.

Schrödinger, E. (1944). *What is life?* Cambridge University Press.

Searle, J. R. (1980). Minds, brains, and programs. *Behavioral and Brain Sciences*, *3*, 417–424.

Searle, J. R. (1990). Is the brain a digital computer? *Proceedings and Addresses of the American Philosophical Association*, *64*(3), 21–37.

Searle, J. R. (2004). *Mind: A brief introduction*. Oxford University Press.

Sellars, W. (1963). *Science, perception, and reality*. Routledge and Kegan Paul.

Shafer-Landau, R. (2012). Evolutionary debunking, moral realism and moral knowledge. *Journal of Ethics & Social Philosophy*, *7*(1), [i]–37.

Shahriar, S. (2022). GAN computers generate arts? A survey on visual arts, music, and literary text generation using generative adversarial network. *Displays*, *73*, 102237.

Shalev-Shwartz, S., & Ben-David, S. (2014). *Understanding machine learning: From theory to algorithms*. Cambridge University Press.

Shannon, C. E. (1948). A mathematical theory of communication. *Bell System Technical Journal*, *27*(3), 379–423.

Shapiro, L. A. (2011). *Embodied cognition*. Routledge.

Shapiro, S. (2003). Vagueness and conversation. In J. C. Beall (Ed.), *Liars and heaps* (pp. 39–72). Oxford University Press.

Shapiro, S. (2006). *Vagueness in context*. Oxford University Press.

Shapiro, S. (2010). Vagueness, metaphysics, and objectivity. In R. Dietz & S. Moruzzi (Eds.), *Cuts and Clouds* (pp. 149–162). Oxford University Press.

Sharma, J., Angelucci, A., & Sur, M. (2000). Induction of visual orientation modules in auditory cortex. *Nature*, *404*(6780), 841–847.

Shermer, M. (2015). Did humans evolve to see things as they really are? *Scientific American*, *313*(5).

Sider, T. (1993). Van Inwagen and the possibility of gunk. *Analysis*, *53*, 285–289.

Sider, T. (2001). Maximality and Intrinsic Properties. *Philosophy and Phenomenological Research*, *63*(2), 357–364.

Sider, T. (2004). Replies to Gallois, Hirsch and Markosian. *Philosophy and Phenomenological Research*, *68*(3), 674–687.

Sider, T. (2011). *Writing the book of the world*. Oxford University Press.

Sider, T. (2013). Against parthood. In K. Bennett & D. W. Zimmerman (Eds.), *Oxford studies in metaphysics, Volume 8* (pp. 236–293). Oxford University Press.

Simon, H. A. (1996). *Models of my life*. MIT Press.

Skow, B. (2009). Relativity and the moving spotlight. *Journal of Philosophy*, *106*(12), 666–678.

Skrzypulec, B. (2016). Ontology of early visual content. *Philosophical Psychology*, *29*(2), 261–276.

Skrzypulec, B. (2018). Thisness and visual objects. *Review of Philosophy and Psychology*, *9*(1), 17–32.

Smith, B. C. (1996). *On the origin of objects*. MIT Press.

Smith, B. C. (2002). Reply to Dennett. In H. Clapin (Ed.), *Philosophy of mental representation* (pp. 237–265). Oxford University Press.

Smith, B. C. (2019). *The promise of artificial intelligence: Reckoning and judgment*. MIT Press.

Smithson, R. (2019). An idealist critique of naturalism. *Inquiry*, *62*(5), 504–526.

Smolin, L. (2014). Time, laws, and the future of cosmology. *Physics Today*, *67*(3), 38.

Soames, S. (1999). *Understanding Truth*, Oxford University Press.

Spelke, E. S., Vishton, P., & von Hofsten, C. (1995). Object perception, object-directed action, and physical knowledge in infancy. In M. S. Gazzaniga (Ed.), *The cognitive neurosciences* (pp. 165–179). MIT Press.

Spratling, M. W. (2013). Image segmentation using a sparse coding model of cortical area V1. *IEEE Transactions on Image Processing*, *22*(4), 1631–1643.

Sprevak, M. (2010). Computation, individuation, and the received view on representation. *Studies in History and Philosophy of Science Part A*, *41*(3), 260–270.

Stanford, P. K., Humphreys, P., Hawley, K., Ladyman, J., & Ross, D. (2010). Protecting rainforest realism. *Metascience*, *19*(2), 161–185.

Stebbing, L. S. (1937). *Philosophy and the physicists*. Dover Publications.

Stein, C. (1956). Inadmissibility of the usual estimator for the mean of a multivariate normal distribution. In J. Neyman (Ed.), *Proceedings of the Third Berkeley Symposium on Mathematical Statistics and Probability* (Vol. 1, pp. 197–206). University of California Press.

Stix, G. (2013). Big neuroscience: Billions and billions (maybe) to unravel mysteries of the brain. *Scientific American Blog* - February 25, 2013.

Stoet, G. (2017). PsyToolkit: A novel web-based method for running online questionnaires and reaction-time experiments. *Teaching of Psychology*, *44*(1), 24–31.

Strang, N. (2017). Transcendental idealism without tears. In T. Goldschmidt & K. Pearce (Eds.), *Idealism: New essays in metaphysics* (pp. 82–103). Oxford University Press.

Strawson, P. F. (1959). *Individuals: An essay in descriptive metaphysics*. Methuen.

Strawson, P. F. (1974). Imagination and perception. In *Freedom and resentment and other essays*. Methuen.

Strawson, P. F. (1987). Kant's philosophy of mind. In R. L. Gregory (Ed.), *The Oxford companion to the mind* (pp. 405–408). Oxford University Press.

Swanson, L. R. (2016). The predictive processing paradigm has roots in Kant. *Frontiers in Systems Neuroscience*, *10*.

Szentágothai, J. (1978). The neuron network of the cerebral cortex: A functional interpretation. *Proceedings of the Royal Society of London*, *201*, 219–248.

Tallant, J. (2013). Quantitative parsimony and the metaphysics of time: Motivating presentism. *Philosophy and Phenomenological Research*, *87*(3), 688–705.

Taylor, C. (1977). *Hegel*. Cambridge University Press.

Tenenbaum, J. B., Griffiths, T. L., & Kemp, C. (2006). Theory-based Bayesian models of inductive learning and reasoning. *Trends in Cognitive Sciences*, *10*(7), 309–318.

Tenenbaum, J. B., Kemp, C., Griffiths, T. L., & Goodman, N. D. (2011). How to grow a mind: Statistics, structure, and abstraction. *Science*, *331*(6022), 1279–1285.

Thomasson, A. L. (2001). Ontological minimalism. *American Philosophical Quarterly*, *38*(4), 319–331.

Thomasson, A. L. (2007). *Ordinary objects*. Oxford University Press.

Thomasson, A. L. (2015). *Ontology made easy*. Oxford University Press.

Trivers, E. (2007). *Mind in life: Biology, phenomenology, and the sciences of mind*. The Belknap Press of Harvard University Press.

Trivers, R. (2011). *The folly of fools: The logic of deceit and self-deception in human life*. Basic Books.

Turing, A. (1950). Computing machinery and intelligence. *Mind*, *59*(236), 433–460.

Tye, M. (1994). Sorites paradoxes and the semantics of vagueness. *Philosophical Perspectives*, *8*, 189–206.

Ullman, S. (1989). Aligning pictorial descriptions: An approach to object recognition. *Cognition*, *32*(3), 193–254.

Unger, P. (1979). There are no ordinary things. *Synthese*, *41*(2), 117–154.

Unger, P. (1980). The problem of the many. *Midwest Studies in Philosophy*, *5*(1), 411–468.

Uzquiano, G. (2004). Plurals and simples. *The Monist*, *87*(3), 429–451.

van Fraassen, B. (1999). The manifest image and the scientific image. In D. Aerts, J. Broekaert, & E. Mathijs (Eds.), *Einstein meets Magritte: An interdisciplinary reflection*. Springer Netherlands.

van Inwagen, P. (1990). *Material beings*. Cornell.

van Inwagen, P. (1998). Meta-ontology. *Erkenntnis*, *48*(2–3), 233–250.

van Inwagen, P. (2009). Being, existence, and ontological commitment. In D. Chalmers, D. Manley, & R. Wasserman (Eds.), *Metametaphysics: New essays on the foundations of ontology*. Oxford University Press.

van Inwagen, P. (2014). *Existence: Essays in ontology*. Cambridge University Press.

Varela, F., Thompson, E., & Rosch, E. (1991). *The embodied mind: Cognitive science and human experience*. MIT Press.

Veatch, H. B. (1978). Is Quine a metaphysician? *The Review of Metaphysics*, *31*(3), 406–430.

Wasserman, R. (2009). Material constitution. In *The Stanford encyclopedia of philosophy* (Winter 2016). Metaphysics Research Lab, Stanford University. Retrieved September 27, 2022, from https://plato.sydney.edu.au/archives/win2016/entries/material-constitution/

Weinberg, J. M., Nichols, S., & Stich, S. (2001). Normativity and epistemic intuitions. *Philosophical Topics*, *29*(1/2), 429–460.

Wertheimer, M. (1912). Experimentelle Studien über das Sehen von Bewegung. *Zeitschrift fur Psychologie*, *61*, 161–165.

Wertheimer, M. (1923). *Untersuchungen zur Lehre von der Gestalt: I*. 2017 Edition. Springer.

Wertheimer, M. (1944). Gestalt theory. *Social Research*, *11*(1), 81–99.

Wertheimer, M. (2012). *A brief history of psychology*. Taylor & Francis Group.

Wheeler, S. C. (1979). On that which is not. *Synthese*, *41*(2), 155–173.

Wiese, W., & Metzinger, T. (2017). Vanilla PP for philosophers: A primer on predictive processing. In T. Metzinger & W. Wiese (Eds.), *Philosophy and predictive processing* (p. 18). MIND Group.

Wiggins, D. (1968). On being in the same place at the same time. *The Philosophical Review*, *77*(1), 90–95.

Wiggins, D. (2001). *Sameness and substance renewed*. Cambridge University Press.

Wigner, E. P. (1967). Remarks on the mind-body question. In J. Mehra (Ed.), *Philosophical Reflections and Syntheses* (pp. 247–260). Springer.
Williams, D. (2018a). Predictive processing and the representation wars. *Minds and Machines*, *28*(1), 141–172.
Williams, D. (2018b). Predictive processing and the representation wars. *Minds and Machines*, *28*(1), 141–172.
Williamson, T. (1994). *Vagueness*. Routledge.
Williamson, T. (2002). *Knowledge and its Limits*. Oxford University Press.
Williamson, T. (2007). *The philosophy of philosophy*. Routledge.
Willshaw, D. J., Buneman, O. P., & Longuet-Higgins, H. C. (1969). Non-holographic associative memory. *Nature*, *222*(5197), 960–962.
Wittgenstein, L. (1921). *Tractatus Logico-Philosophicus*. (D. F. Pears & B. F. McGuinness, Trans.). Humanities Press.
Wittgenstein, L. (1953). *Philosophical investigations*. (G.E.M. Anscombe, Trans.). Blackwell.
Wittgenstein, L. (1965). A lecture on ethics, *Philosophical Review*, *74*(1), 3–12.
Wolfram, S. (2002). *A new kind of science*. Wolfram Media.
Wolpert, D. H., & Macready, W. G. (1997). No free lunch theorems for optimization. *IEEE Transactions on Evolutionary Computation*, *1*(1), 67–82.
Wright, C. (1976). Language-mastery and the sorites paradox. In G. Evans & J. McDowell (Eds.), *Truth and meaning: Essays in semantics*. Clarendon Press.
Wright, C. (1987). Further reflections on the sorites paradox. *Philosophical Topics*, *15*(1), 227–290.
Zadeh, L. A. (1975). Fuzzy logic and approximate reasoning. *Synthese*, *30*(3–4), 407–428.
Zahavi, D. (2018). Brain, mind, world: Predictive coding, neo-Kantianism, and transcendental idealism. *Husserl Studies*, *34*(1), 47–61.
Zhang, H., Fritts, J. E., & Goldman, S. A. (2008). Image segmentation evaluation: A survey of unsupervised methods. *Computer Vision and Image Understanding*, *110*(2), 260–280.
Zimmerman, D. W. (1996). Could extended objects be made out of simple parts? An argument for "Atomless Gunk." *Philosophy and Phenomenological Research*, *56*(1), 1–29.

Index

A. C. Schwaninger, *Cognitive Metaphysics*, Synthese Library 529,
https://doi.org/10.1007/978-3-032-05420-3